T0140535

Springer Tracts in Mechanical Engineering

Series Editors

Seung-Bok Choi, College of Engineering, Inha University, Incheon, Korea (Republic of)

Haibin Duan, Beijing University of Aeronautics and Astronautics, Beijing, China

Yili Fu, Harbin Institute of Technology, Harbin, China

Carlos Guardiola, CMT-Motores Termicos, Polytechnic University of Valencia, Valencia, Spain

Jian-Qiao Sun, University of California, Merced, CA, USA

Young W. Kwon, Naval Postgraduate School, Monterey, CA, USA

Springer Tracts in Mechanical Engineering (STME) publishes the latest developments in Mechanical Engineering - quickly, informally and with high quality. The intent is to cover all the main branches of mechanical engineering, both theoretical and applied, including:

- Engineering Design
- Machinery and Machine Elements
- Mechanical structures and Stress Analysis
- Automotive Engineering
- Engine Technology
- Aerospace Technology and Astronautics
- Nanotechnology and Microengineering
- Control, Robotics, Mechatronics
- MEMS
- Theoretical and Applied Mechanics
- Dynamical Systems, Control
- Fluids mechanics
- Engineering Thermodynamics, Heat and Mass Transfer
- Manufacturing
- Precision engineering, Instrumentation, Measurement
- Materials Engineering
- Tribology and surface technology

Within the scopes of the series are monographs, professional books or graduate textbooks, edited volumes as well as outstanding PhD theses and books purposely devoted to support education in mechanical engineering at graduate and post-graduate levels.

Indexed by SCOPUS and Springerlink. The books of the series are submitted for indexing to Web of Science.

To submit a proposal or request further information, please contact: Dr. Leontina Di Cecco Leontina.dicecco@springer.com or Li Shen Li.shen@springer.com.

Please check our Lecture Notes in Mechanical Engineering at http://www.springer.com/series/11236 if you are interested in conference proceedings. To submit a proposal, please contact Leontina.dicecco@springer.com and Li.shen@springer.com.

More information about this series at http://www.springer.com/series/11693

Sergey Kozinov · Volodymyr Loboda

Fracture Mechanics of Electrically Passive and Active Composites with Periodic Cracking along the Interface

Springer

Sergey Kozinov
Institute of Mechanics and Fluid Dynamics
TU Bergakademie Freiberg
Freiberg, Sachsen, Germany

Volodymyr Loboda
Department of Theoretical
and Computational Mechanics
Oles Honchar Dnipro National University
Dnipro, Ukraine

ISSN 2195-9862   ISSN 2195-9870 (electronic)
Springer Tracts in Mechanical Engineering
ISBN 978-3-030-43140-2   ISBN 978-3-030-43138-9 (eBook)
https://doi.org/10.1007/978-3-030-43138-9

This Springer imprint is published by the registered company Springer Nature Switzerland AG
The registered company address is: Gewerbestrasse 11, 6330 Cham, Switzerland

Preface

Composite materials and adhesive or bonded joints are actively used in wide range of engineering field. In process of such joints manufacture and exploitation numerous defects can occur at the material interface. In some cases, these defects are distributed periodically or they can be approximately considered as periodical. Because the fracture is usually originated from such defects, the problem of the periodic set of the interface cracks is quite important for the applications.

The monograph provides a comprehensive study of the fracture behaviour of the bimaterial composites consisting of the periodically connected components or the bimaterial composites possessing the periodical interface cracks. The monograph is divided in four main parts. The first part is Introduction and Literature overview. The next three parts correspond to the isotropic, anisotropic and piezoelectric/dielectric properties of the bimaterial components, thus, gradually increasing difficulty of the solutions and finally coming to the coupled electromechanical problems. For the case of isotropic and anisotropic materials, the problem for an arbitrary set of cracks were additionally considered, while for the piezoelectric materials the main consideration was devoted to the study of the electric permittivity of the cracks filler implying not only a simple electrically permeable model, but also a physically realistic electrically semi-permeable model. Throughout the monograph, the cracks possessing zones of their faces contact are considered, thus excluding the physically unrealistic interpenetration of the composite components inherent to the classical open model. For the piezoelectric bimaterial, some results are also obtained within the classical model since even they were not available in the literature. A case of possible very large crack faces contact, including the completely closed cracks, is considered as well, allowing the study of the composite behaviour under compression. As a result of the research presented in the monograph, a reader can get better understanding of the fracture processes taking place in the composite materials and start to be acquainted with mathematical methods of complex function theory leading to closed-form solutions. In all parts, the corresponding mechanical or electromechanical fields, stress and electric intensity factors are derived and studied in detail. Consequently, the reader can use mathematical formulae or interpret graphs and data in tables.

This monograph addresses academics and engineers working on isotropic, anisotropic and piezoelectric materials and components, especially those who are active in the analysis of strength and durability of composite constructions. Also, this book is useful for postgraduate students and researchers dealing with the fracture mechanics of composite materials and teaching at the departments of mechanical engineering, civil engineering, material science, electrical engineering and computational engineering. The level of the monograph pre-requisites that the reader is familiar with fracture mechanics and has already some knowledge in analytical methods based on complex functions.

We are grateful for the support of Oles Honchar Dnipro National University, Ukraine, which was significant at all stages of the monograph preparation.

We thank the TU Bergakademie Freiberg, Germany, the German Research Foundation (Deutsche Forschungsgemeinschaft) and personally Prof. M. Kuna for continued support over many years.

Freiberg, Germany   Sergey Kozinov
Dnipro, Ukraine   Volodymyr Loboda
January 2020

Contents

Acronyms

Geometry

h	Period
l	Length of the crack (including the contact zone if present)
M	Open crack faces
L	Contact zones
U	Bonded parts of the interface
b	Point between the contact zone and the open crack face
a	Point between the bonded part of the interface and the contact zone
c	Point between the bonded part of the interface and the open crack face (d in case of arbitrary set of interface cracks, where other crack tips are denoted by $c_j(j = 1..., J - 1)$)
x, y	Cartesian coordinates (cases of isotropic/anisotropic materials)
x_i	Cartesian coordinates (case of piezoelectric materials)

Material Properties

$k = 1$ is related to the "upper" material and $k = 2$ to the "lower" material

E_k	Young's modulus of the isotropic material
ν_k	Poisson's ratio of the isotropic material
μ_k	Shear modulus of the isotropic material
κ_k	$= 3 - 4\nu_k$ for plane strain, $= (3 - \nu_k)/(1 + \nu_k)$ for plane stress
γ	$= \frac{\kappa_1 \mu_2 + \mu_1}{\kappa_2 \mu_1 + \mu_2}$
ε	$= (\ln\gamma)/2\pi$
n''	Bimaterial material constant (case of anisotropic materials)
m	Bimaterial material constant (case of piezoelectric materials)

$s_{ij}^{(k)}$	Compliance coefficients
$C_{ijmq}^{(k)}$	Elastic constants
$e_{imq}^{(k)}$	Piezoelectric constants
$\varepsilon_{iq}^{(k)}$	Dielectric constants

Electromechanical Fields

u_i	Elastic displacements
ε_{ij}	Elastic strains
σ_{ij}	Elastic stresses
ϕ	Electric potential
D_i	Electric displacements
D	Electric displacement inside the crack with limited electric permittivity
$\langle...\rangle$	Jump of some function (property) when crossing the interface from the upper part ($k = 1$) to the lower one ($k = 2$)
λ	Relative contact zone length
K_1, K_2	Stress intensity factors (superscript *osc* is added when considering classical "oscillating" model)
K_4	Electric displacement intensity factor (superscript *osc* is added when considering classical "oscillating" model)
K^*	Dimensionless (normalized) stress intensity factor K_2
G	Energy release rate

Applied Loading

σ	Uniform normal mechanical loading (perpendicular to the interface)
τ	Uniform shear mechanical loading
p	Resulting mechanical loading
β	Angle between the vector of the resulting mechanical loading p and the vertical axis
d	Uniform electric displacement
$\sigma_{11}^{(k)\infty}$	Uniform tensile stresses which provide continuity of displacements u_1 at infinity
$d_1^{(k)\infty}$	Uniform electric displacement which provide continuity of displacements u_1 at infinity

Mathematical Symbols

ln   Natural logarithm
e or exp   Exponential function
$i = \sqrt{-1}$ Imaginary unit
$\Xi(\vartheta)$ $= \sin(\pi\vartheta/h)$

Functions of Complex Variable

z Complex variable ($z = x + iy$ in case of isotropic materials, $z = x + \mu y$ in case of anisotropic materials and $z = x_1 + \rho x_3$ in case of piezoelectric materials)
Re   Real part
Im   Imaginary part
$S(z)$ Canonical solution of periodic combined Dirichlet-Riemann BVP
$Z(z)$ Canonical solution of a periodic Riemann problem
$\psi(z)$ Solution of a periodic Dirichlet problem
$F(z)$ General solution of homogeneous combined Dirichlet-Riemann BVP

Abbreviations

BVP   Boundary value problem
ERR   Energy release rate
IF   Intensity factor
LEFM   Linear elastic fracture mechanics
SIF   Stress intensity factor

Chapter 1
Literature Review on Cracks Located at the Interface of Dissimilar Materials (Interface Cracks)

There is a significant breakthrough in use of composite materials in modern technology, construction, mechanical engineering and electronics in the past decade. Such popularity takes place due to ability to achieve required mechanical properties of the composite. Analyzing stresses and strains in the elements of mechanical structures, the most important issue is to determine stress concentrations near the defects or cracks, which are characterized by stress intensity factors (SIFs). Based on the experimental investigations, fracture of composites is found to be primary caused by the cracks located at the interface of the composite components. Therefore, urgent problem is the development of the special techniques for solving problems of the strength and crack growth resistance taking into account the dissimilarity of the composite components and a discontinuous change of the (electro-)mechanical properties of the materials. Initially, isotropic materials were used as components of the composites. Later on, they were replaced by anisotropic materials. Recently, with the development of smart and active structures, piezoelectric materials started to play an important role. In case of anisotropic materials, fracture analysis is complicated by the appearance of additional bimaterial characteristics such as a degree of the anisotropy and the inclination angle of anisotropy's principal axes to the interface of materials. Furthermore, piezoelectric materials are electrically active with the direct and inverse electromechanical coupling effect. Large amount of research on interface cracks indicates the relevance of their study and further development of effective methods and algorithms. The basic methods for solving problems of internal and interface cracks can be found in known monographs of Muskhelishvili [1], Panasyuk [2], Parton and Kudryavtsev [3], Cherepanov [4], Cruse [5], Hahn [6], Sneddon and Lowengrub [7], Sih [8], Kassir and Sih [9], Altenbach et al. [10], Qin [11], Kienzler [12], Atluri [13], Schwalbe et al. [14], Freund [15], Gdoutos [16], Kanninen and Popelar [17]. The most important results related to the stress intensity factors and other branches of fracture mechanics are summarized in the handbooks Tada et al. [18], Murakami [19] and others.

© The Editor(s) (if applicable) and The Author(s), under exclusive license
to Springer Nature Switzerland AG 2020
S. Kozinov and V. Loboda, *Fracture Mechanics of Electrically Passive
and Active Composites with Periodic Cracking along the Interface*, Springer Tracts
in Mechanical Engineering, https://doi.org/10.1007/978-3-030-43138-9_1

1.1   Classic Model of the Crack Between Two Materials

Isotropic bimaterial. Historically, the first model of the crack between two materials is the model with not contacting faces. First such problem for the crack, located at the interface of two dissimilar isotropic half-planes, under the tensile loading uniformly distributed at infinity, is formulated and solved by Williams [20]. The exact analytical solution is obtained in this paper and it is shown that the oscillation of displacements and stresses occur in the vicinity of the crack tip, namely physically impossible phenomenon of the interpenetration of crack faces into each other is determined. Later on, the plane problems under different boundary conditions at the crack faces and under various kinds of remote loading within this model are solved by Rice and Sih [21], England [22], Erdogan [23]. Methods of integral transformations and complex variable theory were used in these papers. The obtained solutions are accurate. The concept of stress intensity factors of a special kind is introduced in these papers as well. The exact solution to the problem of the penny-shaped crack located at the interface of two different isotropic half-spaces under the remote uniform axially symmetric tensile loading is obtained in the paper of Mossakovsky and Rybka [24]. Complex potential method is used to derive the solution. Axisymmetric problems for a penny shaped interface crack between two isotropic materials were considered also in [25–28]. Elastic fracture mechanics concepts are reexamined for a crack at the interface between dissimilar solids in [29]. The complex stress intensity factor K associated with the elastic interface crack is discussed and, specifically, its validity as a crack tip characterizing parameter is noted for the cases of small scale nonlinear material behavior and/or small scale contact zones at the crack tip. Some possible definitions of the stress intensity factors K_I and K_{II} of classical type associated with interface cracks are discussed.

Anisotropic bimaterial. The earliest work, in which a set of cracks located between anisotropic half-spaces is investigated, is the paper of Clements [30]. The exact analytical solution within the "open" model under the arbitrary mechanical loading at crack faces is presented in this article. The case when the stress-strain state of a body is independent on one of the coordinates is considered. In such case, equations of elasticity theory for an anisotropic body are presented by analytical functions of complex variable. A conjunction problem is obtained, a solution of which is found in the closed form. It should be noted that the papers of Hwu [31], Kattis [32], Qian and Sun [33] are among other studies of the interface crack at the interface of anisotropic materials. Ting [34–36] investigates the oscillation degree of stress and displacement fields for a crack between two anisotropic materials and, in particular, comes to the conclusion that the oscillation degree didn't change at fixed relative orientation of the principal axes of anisotropy.

1.2 Contact Model of the Crack Between Two Materials

The classical model has its advantages and disadvantages. Among advantages is relative simplicity of the problem solving, if compared with the contact model. Among disadvantages are physically unrealistic oscillating singularities in the fields of stresses and displacements in the vicinity of the crack tips and complex stress intensity factors.

In the model, proposed by Comninou [37], a contact of the crack faces in the vicinity of their tips is assumed, besides the values of the contact zones are a priori unknown. In the papers [37, 38] the problem for the cracks between dissimilar isotropic half-planes with smooth contact zones in the vicinity of the crack tips is reduced to an integral equation which is numerically investigated. The contact zone length is uniquely defined by additional equations, derived from the conditions for the stresses and displacements.

Numerical solutions for an isotropic bimaterial. In the papers of Dundurs and Comninou [39], Comninou and Dundurs [40], Ni and Nemat-Nasser [41, 42], Huang et al. [43] the plane problems for cracks with contact zones are reduced to the singular integral equations which are numerically solved. When solving the integral equations numerically, there is a problem with determination of the contact zone length since the integration limits are unknown and must be found from additional equations. This additional conditions are formulated on the basis of the stresses limiting at the points of the crack faces closing.

Since this is a rather complex numerical problem, the attempts are made to obtain analytical solutions.

Analytical solutions for an isotropic bimaterial. The problem for the crack, located between two isotropic materials, in view of one contact zone is analytically solved for the first time by Simonov [46]. The transcendental equation is obtained as a result of the exact solution using complex potentials. The contact zone length of the crack faces is determined from this equation. It is found that the mechanical fields at the crack tips have a square root singularity and the traditionally introduced SIFs are real numbers. Later, analytical solutions for a crack with one or two contact zones are obtained using different methods and under the different external loading in the papers of Simonov [44–46], Beom and Atluri [47], Atkinson [48, 49], Gautesen and Dundurs [50, 51], Dundurs and Gautesen [52], Gautesen [53, 54], Loboda [55]. In these works the influence of one contact zone on another is studied. The conclusions are drawn that neglecting the contact zone near one of the tips leads to inessential errors in determining of the SIF near the other tip. The problem for a set of interface cracks with smooth contact zones is analytically solved in an elastic and thermoelastic cases in the papers of Kharun and Loboda [56, 57], respectively. Arbitrary applied concentrated loading is considered in the paper of Kharun and Loboda [58].

Analytical solutions for an anisotropic bimaterial. A crack at the interface of two anisotropic materials is studied by Wang and Choi [59]. The problem is reduced to a singular integral equation which is solved numerically. A penny-shaped crack with a contact of the crack faces between anisotropic half-spaces is considered in the paper of Qu and Xue [60]. Developing methods of complex function theory, used by Nahmeyn and Nuller [61–64] for mixed problems of rigid stamps, the elastic problem for interface cracks with a contact zone for anisotropic materials is analytically solved in the paper of Herrmann and Loboda [65]. More general case taking into account temperature fields is considered in the paper of Herrmann and Loboda [66]. Rather simple explicit expressions for the stresses along the bonded regions of the materials, expressions for the displacement jumps of crack faces, equations for the determination of the contact zone length are obtained in [65, 66]. Thermoelastic problem for a set of cracks with smooth contact zones is analytically solved in the paper of Kharun and Loboda [67].

1.3 Cracks in Piezoelectric Materials

The fundamentals of mechanics of piezoelectric materials are described in the monographs of Grinchenko et al. [68], Parton and Kudryavtsev [3]. Many obtained results related to this subject were reflected in the review papers [69–72]. In particular, the basic relations of the linear deformation theory of piezoelectric media are considered and a number of problems of electroelasticity are formulated and solved. When solving problems for piezoelectric bodies with cracks, the formulation of electrical conditions at crack faces should be well defined. Cracks are usually filled with some medium (air, water, etc.), so the properties of this medium should be taken into account. Since it is not easy, the special approaches are used to simulate approximate boundary cases, the basic of which are electrically permeable and electrically insulating cracks.

Case of electrically permeable cracks. One of the first papers, dedicated to the study of cracks at the interface of piezoelectric materials, is an article of Kudryavtsev et al. [73]. In this paper the problem for a single crack at the interface of two piezoelectrics is analytically solved within the limiting case of fully electrically permeable cracks. A Green's function for the composite piezoceramics plane with an electrically permeable interface crack is built in the paper of Filshtynsky and Filshtynsky [74]. Analysis of interface cracks in piezoelectric bimaterials in the framework of the electrically permeable crack model is also done in the papers of Wang and Han [75], Gao and Wang [76], Beom [77], Gao et al. [78], Zhou and Wang [79].

Case of electrically insulating cracks. The study of the opposite limiting case of the electrically insulating crack is initiated in the thesis of Deeg [80]. In the paper of Sosa [81] an exact solution for a crack between dissimilar piezoelectric half-planes within the classical model is constructed. The functions similar to the function of Airy are introduced and a solution is presented through analytical functions of a complex variable as a limiting case of the problem for an elliptical hole in an infinite plane. One of the thorough research of the electrically insulating crack is the paper of Suo et al. [82]. In this paper an exact analytical solution for the case of the plane and out-of-plane strain is constructed using the approach of Lekhnitsky [83] and Stroh [84]. The authors introduce the intensity factors of a special kind, analyzing the possible types of power singularities of functions of stresses and displacements in the vicinity of the crack tip. After considering an elliptical hole at a uniform piezoelectric material and a crack as a limiting case of the elliptical hole, Gao and Fan [85] examined the correctness of the simplified electrical conditions for the crack. Since the singularity of the electric displacement, which depends on the electrical loading, appears in the vicinity of the tip of the electrically insulating crack, the authors have concluded that the model of the electrically permeable crack is more realistic.

Methods of complex function theory and theory of integral transforms are used in the papers of Park [86], Ru et al. [87], Ru [88], Shen et al. [89], Wang [90], Wang and Meguid [91], in which cracks between dissimilar piezoelectric materials are studied in various limiting cases of the electric permittivity of the crack faces.

Modeling of finite permittivity of cracks. Conditions on the crack faces that allow to take into account the properties of the crack filler are offered by Parton and Kudryavtsev [3] and Hao and Shen [92]. This is the approach for taking into account the finite electric permittivity of a crack. In this model the crack is considered to be a capacitor and the boundary conditions of a capacitor are used. This approach has attracted considerable attention of researchers. Among them the papers of Dunn [93], McMeeking [94], Xu and Rajapakse [95], Wang and May [96] should be noted. The problem is solved by the finite element method in the paper of Gruebner et al. [97]. A homogeneous piezoelectric material is considered in all the above articles. In the paper of Govorukha et al. [98] the crack at the interface of two different piezoelectric materials is analytically investigated taking into account the electric permittivity of the crack filler. Additionally, the crack in a homogeneous piezoelectric material is analyzed as a special case. Li and Chen [99] investigated a crack with finite electric permittivity located at the interface of piezoelectric and dielectric materials. Energetically consistent boundary conditions for cracks in homogeneous materials are proposed in the paper of Landis [100] and are developed by Li et al. [101], Ricoeur and Kuna [102].

Study of cracks in piezoelectric bimaterials within the contact model. The solution of a problem for a crack in view of a contact zone is complicated. Therefore, such studies have been started only recently. A thermoelastic problem is investigated in the paper of Qin and Mai [103] within the specified model by a method of singular integral equations. It should be mentioned that in the papers of Herrmann and Loboda [104] and Herrmann et al. [105] the exact analytical solution is obtained for electrically permeable and electrically insulating cracks located at the interface of two piezoelectric materials.

1.4 Modeling of Cracks with Friction in the Contact Zone and Other Ways of Modeling of Pre-fracture Zones Near the Crack Tips

Interface cracks with contact zones in isotropic bimaterials accounting for friction are investigated in the papers [106–108] and also in an analytical way in the works of Antipov [109], Ostrik [110], Sapsathiarn et al. [111]. The given problems are reduced to integral equations, for which approximate asymptotic solutions are obtained. The problem for a crack, located at the interface of orthotropic materials, taking into account the friction in the contact zone, is analytically solved in the paper of Loboda and Kharun [112]. Another way of modeling of zones near the tip of the interface crack is proposed in the papers of Kaminsky et al. [113, 114], Kaminsky et al. [115–117], Loboda and Sheveleva [118]. In these papers, zones of weakened interparticle bonds are introduced on the extension of the crack.

1.5 Periodic Sets of Cracks

Connection of two dissimilar materials is often made by periodical jointing while other parts of the interface remain free. These parts can be considered as cracks. For example, the welding of materials is usually done with a certain step, so such connections are periodic. Another case when the study of a periodic set is reasonable, is the case of finite number of cracks. If cracks have approximately the same length, the distance between them is roughly the same and the number of cracks is large, it is more convenient to use the solution for a periodic set of cracks than investigate this set of cracks explicitly.

The plane problem for a periodic set of cracks is investigated in the paper of Sulim et al. [119], where the exact analytical solution is obtained in the closed form by solving singular integral equations. In the paper of Nahmeyn et al. [120] deformation of the composed elastic plane weakened by a periodic set of fully open or fully closed cracks is considered.

A periodic set of cracks located at the interface of a piezoelectric and an isotropic conductor is studied within the classical model in the paper of Kudryavtsev and Rakitin [121]. The crack faces are assumed to be loaded by an uniform pressure. A solution is found in the form of series, and expressions for the normal stresses and displacements along the interface of materials are obtained. Kaloyerov and Boronenko [122] investigated a periodic problem of magnetoelasticity for a body with elastic inclusions by complex potential method.

A periodic set of fully electrically permeable cracks at the interface of two piezoelectric materials is considered in the paper of Gao and Wang [76] using the approach of Lekhnitsky [83] and Stroh [84]. Analytical expressions for the intensity factors and electric displacement are obtained. Further investigation is carried out in the paper of Gao et al. [78] where the obtained solution is numerically realized for the special case of a single crack. Häusler et al. [123] solved the problem for collinear and periodic electrode-ceramic interfacial cracks in piezoelectric bimaterials. The expressions for the intensity factors and the energy release rate are obtained, and the results of numerical calculations are provided.

A periodic set of interface cracks between isotropic materials is studied within the contact model in the paper of Schmueser and Comninou [124] by constructing and numerically solving a system of singular integral equations. As a result of the obtained solution, the stress intensity factors are found and their dependence on the mechanical and geometric characteristics of the bimaterial is investigated. Periodic sets of interface cracks in piezoelectric bimaterials were also studied in papers [125, 126].

It should be noted that the periodic set of cracks in isotropic, anisotropic and piezoelectric bimaterial spaces is not studied enough, and its investigation has a significant scientific interest. The study of various electrical conditions at the crack faces in a piezoelectric bimaterial has a particular interest as well, especially the models of the electrically permeable crack and the crack with the finite electric permittivity.

References

1. Muskhelishvili N (1977) Some basic problems of the mathematical theory of elasticity. Springer, Dordrecht
2. Panasyuk V (1968) Limit equilibrium of brittle bodies with cracks. Naukova Dumka, Kyiv (translation in English: Michigan information service, Detroit, 1971) (in Russian)
3. Parton V, Kudryavtsev B (1988) Electromagnetoelasticity. Gordon and Breach Science Publishers, New York
4. Cherepanov G (1979) Mechanics of brittle fracture. McGraw-Hill International Book Co., New York
5. Cruse T (1988) Boundary element analysis in computational fracture mechanics. Kluwer Academic Publishers, Dordrecht
6. Hahn H (1976) Bruchmechanik: Einführung in die theoretischen Grundlagen. Mechanik, Teubner-Studienbüche, Stuttgart
7. Sneddon L, Lowengrub M (1969) Crack problems in the classical theory of elasticity. Wiley, New York
8. Sih G (1973) Methods of analysis and solutions of crack problems. Mechanics of fracture, vol 1. Noordhoff International Publisher, Leyden
9. Kassir M, Sih G (1975) Three dimensional crack problems. Mechanics of fracture, vol 2. Noordhoff International Publisher, Leyden
10. Altenbach H, Altenbach J, Rikards R (1996) Einführung in die Mechanik der Laminatwerkstoffe. Deutscher Verlag für Grundstoffindustrie, Stuttgart
11. Qin Q (2001) Fracture mechanics of piezoelectric materials. WIT Press, Southampton and Boston
12. Kienzler R (1993) Konzepte der Bruchmechanik. Vieweg, Wiesbaden
13. Atluri S (1986) Computational methods in the mechanics of fracture. Elsevier Science Publisher, Noorth-Holland
14. Schwalbe K, Scheider I, Cornec A (2013) Guidelines for applying cohesive models to the damage behaviour of engineering materials. Springer, Heidelberg
15. Freund L (1990) Dynamic fracture mechanics. Cambridge University Press, Cambridge
16. Gdoutos E (1990) Fracture mechanics criteria and applications. Kluwer, Dordrecht, The Netherlands

17. Kanninen M, Popelar C (1985) Advanced fracture mechanics. Oxford University Press, New York
18. Tada H, Paris P, Irwin G (1985) The stress analysis of cracks handbook, 2nd edn. Paris Production Inc., St. Louis
19. Murakami Y (1987) Stress intensity factors handbook, vols 1–5. Pergamon Press, Oxford
20. Williams ML (1959) The stresses around a fault or cracks in dissimilar media. Bull Seism Soc America 49:199–204
21. Rice JR, Sih GC (1965) Plane problems of cracks in dissimilar media. J Appl Mech 32:418–423
22. England A (1965) A crack between dissimilar media. Trans ASME J Appl Mech 32:400–402
23. Erdogan F (1965) Stress distribution in bonded dissimilar materials with cracks. Trans ASME J Appl Mech 32(15):2027–2040
24. Mossakovsky V, Rybka M (1964) Generalization of the Griffith-Sneddon criterion for the case of a nonhomogeneous body. J Appl Math Mech 28(6):1277–1286
25. Nahta R, Moran B (1993) Domain integrals for axisymmetric interface crack problems. Int J Solids Struct 30:403–410
26. Martin-Moran C, Barber J, Comninou M (1983) The penny-shaped interface crack with heat flow. Part 1: perfect contact. J Appl Mech 50:29–36
27. Martin-Moran C, Barber J, Comninou M (1983) The penny-shaped interface crack with heat flow. Part 2: imperfect contact. J Appl Mech 50:770–776
28. Zhao M, Dang H, Fan C, Chen Z (2016) Analysis of an arbitrarily shaped interface cracks in a three-dimensional isotropic thermoelastic bi-material. Part 1: theoretical solution. Int J Solids Struct 97:168–181
29. Rice J (1988) Elastic fracture mechanics concept for interfacial cracks. J Appl Mech 55:98–103
30. Clements D (1971) A crack between dissimilar anisotropic media. Int J Engen Sci 9:257–265
31. Hwu C (1993) Fracture parameters for the orthotropic bimaterial interface cracks. Eng Fract Mech 45:89–97
32. Kattis M (1999) Nonplanar interfacial cracks in anisotropic bimaterials. Int J Fract 98:313–327
33. Quan W, Sun CT (1998) Methods for calculating stress intensity factors for interfacial cracks between two orthotropic solids. Int J Solids Struct 35:3317–3330
34. Ting TCT (1986) Explicit solution and invariance of the singularities at an interface crack in anisotropic composites. Int J Solids Struct 22:965–983
35. Ting TCT (1990) Interface cracks in anisotropic bimaterial. J Mech Phys Solids 38:505–513
36. Ting TCT (2000) Recent developments in anisotropic elasticity. Int J Solids Struct 37:401–409
37. Comninou M (1977) The interface crack. J Appl Mech 44:631–636
38. Comninou M (1978) The interface crack in a shear field. ASME J Appl Mech 45:287–290
39. Dundurs J, Comninou M (1979) Some consequences of inequality conditions in contact and crack problems. J Elast 9:71–82
40. Comninou M, Dundurs J (1983) Partial closure of cracks at the interface between a layer and a hald-space. Eng Fract Mech 18:315–323
41. Ni L, Nemat-Nasser S (1991) Interface cracks in anisotropic dissimilar materials: an analytical solution. J Mech Phys Solids 39:113–144
42. Ni L, Nemat-Nasser S (1992) Interface cracks in anisotropic dissimilar materials: general case. Quaterly Appl Math 2:305–322
43. Huang Y, Wang W, Liu C, Rosakis A (1998) Intersonic crack growth in bimaterial interfaces: an investigation of crack face contact. J Mech Phys Solids 46:2233–2259
44. Simonov IV (1984) On the steady motion of a crack with slip and separation sections along the interface of two elastic materials. J Appl Math Mech 48(3):347–353. https://doi.org/10.1016/0021-8928(84)90144-8
45. Simonov I (1985) Brittle cleavage of a piecewise-homogeneous elastic medium. J Appl Math Mech 49(2):207–214
46. Simonov IV (1986) Crack at an interface in a uniform stress field. Mech Compos Mater 21:650–657. https://doi.org/10.1007/BF00605924

47. Beom H, Atluri S (1996) Near-tip fields and intensity factors for interfacial cracks in dissimilar anisotropic piezoelectric media. Int J Fract 75:163–183
48. Atkinson C (1977) On stress singularities and interfaces in linear elastic fracture mechanics. Int J Fract 13:807–820
49. Atkinson C (1982) The interface crack with a contact zone (an analytical treatment). Int J Fract 18:161–177
50. Gautesen A, Dundurs J (1987) The interface crack in a tension field. J Appl Mech 54:93–98
51. Gautesen A, Dundurs J (1988) The interface crack under a combined loading. ASME J Appl Mech 55:580–586
52. Dundurs J, Gautesen A (1988) An opportunistic analysis of the interface crack. Int J Fract 36:151–159
53. Gautesen A (1992) The interface crack in a tension field: an eigenvalue problem for the gap. Int J Fract 55:261–271
54. Gautesen A (1993) The interface crack under a combined loading. Int J Fract 60:349–361
55. Loboda V (1993) The quasi-invariant in the theory of interface cracks. Eng Fract Mech 44:573–580
56. Kharun I, Loboda V (2003) A set of interface cracks with contact zones in combined tension-shear field. Acta Mechanica 166:43–56
57. Kharun I, Loboda V (2004) A thermoelastic problem for interface cracks with contact zones. Int J Solids Struct 41:159–175
58. Kharun I, Loboda V (2002) Interface cracks with contact zones in the field of concentrated forces and moments. Math Methods Phys-Mech Fields 45(2):103–113 (in Ukrainian)
59. Wang SS, Choi I (1983) The interface crack between dissimilar anisotropic composite materials. J Appl Mech 50:169–178
60. Qu J, Xue Y (1998) Three-dimensional interface cracks in anisotropic bimaterials: the non-oscillatory case. J Appl Mech 65:1048–1055
61. Nakhmein E, Nuller B (1976) A method for solving of contact periodic problems for the elastic strip and ring. USSR AS, MTT 40(3):53–61 (in Russian)
62. Nakhmein E, Nuller B (1986) Contact between an elastic half-plane and a partly separated stamp. J Appl Math Mech 50(4):507–515
63. Nakhmein E, Nuller B (1988) The pressure of a system of stamps on an elastic half-plane under general conditions of contact adhesion and slip. J Appl Math Mech 52(2):223–230
64. Nakhmein E, Nuller B (1992) Combined periodic boundary-value problems and their applications in the theory of elasticity. J Appl Math Mech 56:82–89
65. Herrmann K, Loboda V (1999) On interface crack models with contact zones situated in an anisotropic bimaterial. Arch Appl Mech 69:317–335
66. Herrmann K, Loboda V (2001) Contact zones models for an interface crack in a thermomechanically loaded anisotropic bimaterial. J Therm Stress 24:479–506
67. Kharun I, Loboda V (2004) A problem of thermoelasticity for a set of interface cracks with contact zones between dissimilar anisotropic materials. Mech Mater 7:585–600
68. Grinchenko V, Ulitko A, Shulga N (1989) Electroelasticity. In Mechanics of coupled fields in the elements of constructions, 5 vol, Naukova Dumka (in Russian)
69. Zhang T, Zhao M, Tong P (2002) Fracture of piezoelectric ceramics. Adv Appl Mech 38:147–289
70. Chen Y-H, Lu T (2003) Cracks and fracture in piezoelectrics. Adv Appl Mech 39:121–215
71. Zhang T, Gao C-F (2004) Fracture behaviors of piezoelectric materials. Theor Appl Fract Mech 41:339–379
72. Schneider G (2007) Influence of electric field and mechanical stresses on the fracture of ferroelectrics. Ann Rev Mater Res 37:491–538
73. Kudryavtsev B, Parton V, Rakitin V (1975) Fracture mechanics of piezoelectric materials. Rectilinear tunnel crack on the boundary with a conductor. J Appl Math Mech 39(1):136–146
74. Fil'shtinskii L, Fil'shtinskii M (1994) Green's function for a composite piezoceramic plane with a crack between phases. J Appl Math Mech 58(2):355–362
75. Wang T, Han X (1999) Fracture mechanics of piezoelectric materials. Int J Fract 98:15–35

76. Gao C-F, Wang M (2000) Collinear permeable cracks between dissimilar piezoelectric materials. Int J Solids Struct 37:4969–4986
77. Beom H (2003) Permeable cracks between two dissimilar piezoelectric materials. Int J Solids Struct 40:6669–6679
78. Gao C-F, Hausler C, Balke H (2004) Periodic permeable interface cracks in pizoelectric materials. Int J Solids Struct 41:323–335
79. Zhou Z-G, Wang B (2006) Investigation of behavior of Mode-I interface crack in piezoelectric materials by using Schmidt method. Appl Math Mech 27:871–882
80. Deeg W (1980) The analysis of dislocation, crack and inclusion problems in piezoelectric solids. PhD thesis, Stanford University
81. Sosa H (1991) Plane problems in piezoelectric media with defects. Int J Solids Struct 28:491–505
82. Suo Z, Kuo CM, Barnett DM, Willis JR (1992) Fracture mechanics for piezoelectric ceramics. J Mech Phys Solids 40:739–765
83. Lekhnitsky S (1963) Theory of elasticity of an anisotropic elastic body. San Francisco: Holden-Day
84. Stroh AN (1962) Steady state problems in anisotropic elasticity. J Math Phys 41:77–103
85. Gao C-F, Fan W (1999) Exact solution for the plane problem in piezoelectric materials with an elliptic hole or a crack. Int J Solid Struct 36:2527–2540
86. Pak Y (1992) Linear electro-elastic fracture mechanics of piezoelectric materials. Int J Fract 54:79–100
87. Ru CQ, Mao X, Epstein M (1998) Electric-field induced interfacial cracking in multilayer electrostrictive actuators. J Mech Phys Solids 46:1301–1318
88. Ru CQ (1999) Effect of electrical polarization saturation on stress intensity factors in a piezoelectric ceramic. Int J Solids Struct 36:869–883
89. Shen S, Nishioka T, Hu SL (2000) Crack propagation along the interface of piezoelectric bimaterial. Theor Appl Fract Mech 34:185–203
90. Wang XD (2000) Analysis of strip electric saturation model of crack problem in piezoelectric materials. Int J Solids Struct 37:6031–6049
91. Wang XD, Meguid SA (2000) On the electroelastic behaviour of a thin piezoelectric actuator attached to an infinite host structure. Int J Solids Struct 37:3231–3251
92. Hao T, Shen Z (1994) A new electric boundary condition of electric fracture mechanics and its application. Eng Fract Mech 47:793–802
93. Dunn M (1994) The effect of crack faces boundary conditions on the fracture mechanics of piezoelectric solids. Eng Fract Mech 48:25–39
94. McMeeking R (1999) Crack tip energy release rate for a piezoelectric compact tension specimen. Eng Fract Mech 64:217–244
95. Xu X, Rajapakse RKND (2001) On a plane crack in piezoelectric solids. Int J Solids Struct 38:7643–7658
96. Wang BL, May YW (2003) On the electrical boundary conditions on the crack surfaces in piezoelectric ceramics. Int J Eng Sci 41:633–652
97. Gruebner O, Kamlah M, Munz D (2003) Finite element analysis of cracks in piezoelectric materials taking into account the permittivity of the crack medium. Eng Fract Mech 70:1399–1413
98. Govorukha V, Loboda V, Kamlah M (2006) On the influence of the electric permeability on an interface crack in a piezoelectric bimaterial compound. Int J Solid Struct 43:1979–1990
99. Li Q, Chen Y (2008) Solution for a semi-permeable interface crack in elastic dielectric/piezoelectric bimaterials. ASME J Appl Mech 75:1–13
100. Landis C (2004) Electrically consistent boundary conditions for electromechanical fracture. Int J Solids Struct 41:6291–6315
101. Li W, McMeeking R, Landis C (2008) On the crack face boundary conditions in electromechanical fracture and an experiment protocol for determining energy release rates. Eur J Mech A/Solids 27:285–301

102. Ricoeur A, Kuna M (2009) Electrostatic traction at dielectric interfaces and their implication for crack boundary conditions. Mech Res Commun 36:330–335

103. Qin Q, Mai Y-W (1999) A closed crack tip model for interface cracks in thermopiezoelectric materials. Int J Solids Struct 36:2463–2479

104. Herrmann K, Loboda V (2000) Fracture mechanical assessment of electrically permeable interface cracks in piezoelectric bimaterials by consideration of various contact zone models. Arch Appl Mech 70:127–143

105. Herrmann K, Loboda V, Govorukha V (2001) On contact zone model for an interface crack with electrically insulated crack surfaces in a piezoelectric bimaterial. Int J Fract 111:203–227

106. Comninou M (1977) Interface crack with friction in the contact zone. J Appl Mech 44(4):780–781

107. Comninou M, Dundurs J (1980) Effect of friction on the interface crack loaded in shear. J Elast 10(2):203–212

108. Leguillon D (1999) Interface crack tip singularity with contact and friction. Comptes Rendus de l'Académie des Sciences - Series IIB - Mechanics-Physics-Astronomy, vol 327, no 5, pp 437–442

109. Antipov Y (1995) An interface crack between elastic materials when there is dry friction. J Appl Math Mech 59(2):273–287

110. Ostrik V (2003) Friction contact of the edges of an interface crack under the conditions of tension and shear. Mater Sci 39(2):214–224

111. Sapsathiarn Y, Senjuntichai T, Rajapakse R (2012) Cylindrical interface cracks in 1-3 piezo-composites. Compos: Part B 43:2257–2264

112. Loboda V, Kharun I (2001) Plane problem of a crack on the interface of orthotropic plates with friction of crack lips. Mater Sci 37(5):735–745

113. Kaminsky A, Kipnis L, Kolmakova V (1995) Slip lines at the end of a cut at the interface of different media. Int Appl Mech 31(6):491–495

114. Kaminsky A, Kipnis L, Kolmakova V (1999) On the Dugdale model for a crack at the interface of different media. Int Appl Mech 35(1):58–63

115. Kaminsky A, Kipnis L, Dudik I (2004) Initial development of the prefracture zone near the tip of a crack reaching the interface between dissimilar media. Int Appl Mech 40(2):176–182

116. Kaminsky A, Dudik I, Kipnis L (2006) On the direction of development of a thin fracture process zone at the tip of an interfacial crack between dissimilar media. Int Appl Mech 42(2):136–144

117. Kaminsky A, Dudik I, Kipnis L (2007) Initial kinking of an interface crack between two elastic media. Int Appl Mech 43(10):1090–1099

118. Loboda V, Sheveleva A (2003) Determining prefracture zones at a crack tip between two elastic orthotropic bodies. Int Appl Mech 39(5):566–572

119. Sulim G, Grilitskii D, Belokur I (1977) Periodic problem for composite plane with cracks. Mater Sci 13(1):72–75

120. Nakhmein E, Nuller B, Ryvkin M (1982) Deformation of a composite elastic plane weakened by a periodic system of the arbitrarily loaded slits. J Appl Math Mech 45(6):821–826

121. Kudryavtsev B, Rakitin V (1976) Periodic set of cracks at the interface of piezoelectric and solid conductor. USSR Acad Sci Mech Solids 2:121–129 (in Russian)

122. Kaloerov S, Boronenko O (2006) Magnetoelastic problem for a body with periodic elastic inclusions. Int Appl Mech 42(9):989–996

123. Häusler C, Gao C-F, Balke H (2004) Collinear and periodic electrode-ceramic interfacial cracks in piezoelectric bimaterials. ASME J Appl Mech 71:486–492

124. Schmueser D, Comninou M (1979) The periodic array of interface cracks and their interaction. Int J Solids Struct 15:927–934

125. Ru C (2000) Electrode-ceramic interfacial cracks in piezoelectric multilayer materials. Trans ASME J Appl Mech 67:255–261

126. Liu M, Hsia K (2003) Interfacial cracks between piezoelectric and elastic materials under in-plane electric loading. J Mech Phys Solids 51:921–944

Chapter 2
Set of Cracks with Contact Zones Located at the Interface of Two Isotropic Materials

A solution to the problem for a *periodic* set of cracks with contact zones located at the interface of two dissimilar isotropic materials is constructed in a closed form in Sect. 2.1. By presenting mechanical fields through the piecewise analytical functions, the problem is reduced to a homogeneous combined periodic Dirichlet-Riemann boundary value problem, a solution of which is obtained in a closed form. As a result of the numerical analysis of the obtained solution for various combinations of materials, the dependence of the relative contact zone length and the SIF on a ratio of the crack length to the period and on an external loading is investigated.

Additionally, a problem for an infinite plane consisting of two dissimilar isotropic materials with an *arbitrary* set of interface cracks is formulated and its solution is derived. The cracks possess smooth contact zones and are in the field of tensile and shear loading distributed uniformly at infinity (Sect. 2.2). The equations for the determination of the contact zone lengths and the expressions for the SIFs at the crack tips are obtained.

2.1 Periodic Set of Interface Cracks

2.1.1 Statement of the Problem and Its Reduction to the Periodic Dirichlet-Riemann Boundary Value Problem

Statement of the problem. Consider a periodic set of cracks arranged with period h (cf. Fig. 1.1) [1]. The cracks are located along the interface of two dissimilar half-planes or half-spaces. The materials are assumed to be isotropic with Young's moduli and Poisson's ratios denoted by E_k and v_k, respectively, where $k = 1$ is related to the "upper" material and $k = 2$ to the "lower" one. The plane is subjected to a uniform

S. Kozinov and V. Loboda, *Fracture Mechanics of Electrically Passive and Active Composites with Periodic Cracking along the Interface*, Springer Tracts in Mechanical Engineering, https://doi.org/10.1007/978-3-030-43138-9_2

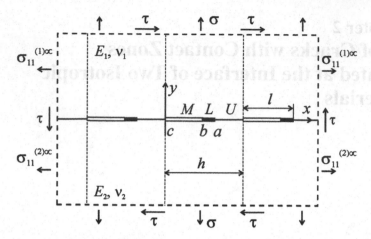

Fig. 2.1 Periodic set of interface cracks between two isotropic materials

tensile (σ) and shear (τ) loading. Furthermore, each half-plane is subjected to the normal stresses $\sigma_{11}^{(k)\infty}$, which provide continuity of displacements u_1 passing through the interface of materials.

The crack faces are assumed to be able to contact without friction at some parts of the cracks. These zones typically occur at the crack tips only. It is known [2] that the contact zones near the interface crack tips are, as a rule, small. It was shown in [3–5] that one of the contact zones has no significant influence on the length of another zone and the corresponding SIF. Usually the error in determination of the contact zone length and the corresponding SIF that arises because of neglecting another contact zone is not exceeding 1%. Therefore, in determination of the SIF at some crack tip only the contact zone that is located near this tip can be taken into account.

Introduce the following denotation: the point between the contact zone and the open crack face is denoted by b, the point between the bonded interface and the contact zone - by a, the point between the bonded interface and the open crack face - by c. The union of the open crack faces will be denoted by M, the contact zones by L and the bonded parts of the interface by U. Assuming the contact zones to be smooth and the open parts of the cracks to be load-free, the continuity and the boundary conditions can be written in Cartesian coordinates xy, depicted in Fig. 2.1, as

$$\langle \sigma_{22}(x) \rangle = 0, \quad \langle \sigma_{12}(x) \rangle = 0, \quad x \in L \cup M \cup U, \tag{2.1}$$

$$\langle u_2(x) \rangle = 0, \quad \langle u_1(x) \rangle = 0, \quad x \in U, \tag{2.2}$$

$$\sigma_{22}^{(1)}(x, 0) = 0, \quad \sigma_{12}^{(1)}(x, 0) = 0, \quad x \in M, \tag{2.3}$$

$$\sigma_{12}^{(1)}(x, 0) = 0, \quad \langle u_2(x) \rangle = 0, \quad x \in L, \tag{2.4}$$

where $\sigma_{ij}^{(k)}$ and $u_i^{(k)}$ are stress and displacement components in the "upper" $(k = 1)$ and the "lower" $(k = 2)$ materials,

$$\langle \sigma_{ij}(x) \rangle = \sigma_{ij}^{(1)}(x, +0) - \sigma_{ij}^{(2)}(x, -0), \quad \langle u_i(x) \rangle = u_i^{(1)}(x, +0) - u_i^{(2)}(x, -0).$$

Construction of the main complex function expressions. The fields of stresses and displacements for a plane problem of elasticity can be expressed via two analytical functions $\varphi_k(z)$ and $\psi_k(z)$ of a complex variable $z = x + iy$ by means of the formulae [6]

$$\begin{cases} \sigma_{22}^{(k)}(x, y) - i\sigma_{12}^{(k)}(x, y) = \Phi_k(z) + \overline{\Phi_k(z)} + z\overline{\Phi_k'(z)} + \overline{\Psi_k(z)}, \\ 2\mu_k[u_1^{(k)}(x, y) + iu_2^{(k)}(x, y)] = \kappa_k\varphi_k(z) - z\overline{\Phi_k(z)} - \overline{\psi_k(z)}, \end{cases} \tag{2.5}$$

$$\sigma_{11}^{(k)}(x, y) + \sigma_{22}^{(k)}(x, y) = 4\operatorname{Re}\Phi_k(z), \tag{2.6}$$

where

$$\Phi_k(z) = \varphi_k'(z),$$

$$\Psi_k(z) = \psi_k'(z),$$

$$\mu_k = \frac{E_k}{2(1+\nu_k)},$$

$$\kappa_k = \begin{cases} 3 - 4\nu_k & \text{- for plane strain,} \\ (3 - \nu_k)/(1 + \nu_k) & \text{- for plane stress.} \end{cases}$$

Note that stresses $\sigma_{11}^{(1)\infty}$ and $\sigma_{11}^{(2)\infty}$ must satisfy the following continuity conditions [7]:

$$\frac{1 + \kappa_1}{\mu_1}\sigma_{11}^{(1)\infty} - \frac{1 + \kappa_2}{\mu_2}\sigma_{11}^{(2)\infty} = \left[\frac{\kappa_2 - 3}{\mu_2} - \frac{\kappa_1 - 3}{\mu_1}\right]\sigma.$$

Introducing the function

$$\omega_k(z) = z\overline{\varphi_k'(z)} + \overline{\psi_k(z)}$$

and denoting $\Omega_k(z) = \omega_k'(z)$, after some manipulations Eq. (2.5) can be rewritten in the following form:

$$\begin{cases} \sigma_{22}^{(k)}(x, y) - i\sigma_{12}^{(k)}(x, y) = \Phi_k(z) + (z - \overline{z})\overline{\Phi_k'(z)} + \Omega_k(\overline{z}), \\ 2\mu_k[u_1^{(k)}(x, y) + iu_2^{(k)}(x, y)] = \kappa_k\varphi_k(z) + (\overline{z} - z)\overline{\Phi_k(z)} - \omega_k(\overline{z}). \end{cases} \tag{2.7}$$

Inserting Eq. (2.7) into boundary conditions (2.1) and (2.2) leads to the following equations for the boundary values of the unknown functions:

$$\begin{cases} \Phi_1^+(x) - \Omega_2^+(x) = \Phi_2^-(x) - \Omega_1^-(x), & x \in L \cup M \cup U, \\ \dfrac{\kappa_1}{2\mu_1}\Phi_1^+(x) + \dfrac{1}{2\mu_2}\Omega_2^+(x) = \dfrac{\kappa_2}{2\mu_2}\Phi_2^-(x) + \dfrac{1}{2\mu_1}\Omega_1^-(x), & x \in U. \end{cases}$$

As it is known [8], if functions which are analytical in some areas have the same values at the boundary, they can be analytically extended. Thus, from the above equalities follows that a function $B(z)$ exists, which is analytical in the entire plane, and a function $A(z)$ exists, which is analytical in the entire plane except for $M \cup L$. They have the form

$$A(z) = \begin{cases} \dfrac{\kappa_1}{2\mu_1}\Phi_1(z) + \dfrac{1}{2\mu_2}\Omega_2(z), & y > 0, \\ \dfrac{\kappa_2}{2\mu_2}\Phi_2(z) + \dfrac{1}{2\mu_1}\Omega_1(z), & y < 0, \end{cases} \qquad B(z) = \begin{cases} \Phi_1(z) - \Omega_2(z), & y > 0, \\ \Phi_2(z) - \Omega_1(z), & y < 0. \end{cases}$$

$$(2.8)$$

Applying Liouville's theorem to the function $B(z)$ and taking into account the behaviour of the functions $\Phi_k(z)$ and $\Omega_k(z)$ at infinity, the expression $B(z) \equiv B = \text{const}$ holds true, and

$$B = \frac{1}{4gs_2}\left(\sigma_{11}^{(1)\infty} + \frac{\gamma - 3}{1+\gamma}\sigma\right),$$

where

$$g = \frac{2\mu_1\mu_2}{\kappa_1\mu_2 + \mu_1}, \quad \gamma = \frac{\kappa_1\mu_2 + \mu_1}{\kappa_2\mu_1 + \mu_2}, \quad s_k = \frac{1 + \kappa_k}{2\mu_k}\frac{\gamma}{1+\gamma}.$$

Resolving Eq. (2.8) with respect to $\Phi_k(z)$ and $\Omega_k(z)$ gives

$$\Phi_k(z) = g\gamma_k^*[A(z) + B/2\mu_m], \quad \Omega_k(z) = g\gamma_m^*[A(z) - \kappa_m B/2\mu_m], \qquad (2.9)$$

where $\gamma_1^* = 1, \; \gamma_2^* = \gamma$.

The subscript $m = 1$, if $k = 2$, and $m = 2$, if $k = 1$.

Introducing a function

$$F(z) = A(z) + \frac{1 - \kappa_1\kappa_2}{2[\mu_1(1 + \kappa_2) + \mu_2(1 + \kappa_1)]}B$$

and using Eq. (2.9), expressions (2.6) and (2.7) take the following form:

$$\sigma_{22}^{(k)} - i\sigma_{12}^{(k)} = g[\gamma_k^* F(z) + \gamma_k^*(z - \bar{z})\overline{F}'(\bar{z}) + \gamma_m^* F(\bar{z})], \tag{2.10}$$

$$\sigma_{22}^{(k)} + \sigma_{11}^{(k)} = 4g\,\mathrm{Re}[\gamma_k^* F(z) + s_m B], \tag{2.11}$$

$$u_1'^{(k)} + iu_2'^{(k)} = \frac{g}{2\mu_k}[\gamma_k^* \kappa_k F(z) - \gamma_k^*(z - \bar{z})\overline{F}'(\bar{z}) - \gamma_k^* F(\bar{z}) + s_m(1 + \kappa_k)B]. \tag{2.12}$$

Thus, fields of stresses and displacements are expressed through a single function $F(z)$ which is analytical in the entire plane except for $L \cup M$. The behaviour of the function $F(z)$ at infinity can be defined using (2.10) as

$$F(z) = \frac{\sigma - i\tau}{g(1 + \gamma)} + O(z^{-2}), \quad z \to \pm i\infty. \tag{2.13}$$

Considering the stress and displacement fields defined by Eq. (2.10)–(2.12) at the interface $y = 0$, the following representation can be written:

$$\sigma_{22}^{(1)}(x, 0) - i\sigma_{12}^{(1)}(x, 0) = g[F^+(x) + \gamma F^-(x)],$$
$$\langle u_1'(x)\rangle + i\langle u_2'(x)\rangle = F^+(x) - F^-(x). \tag{2.14}$$

Relations (2.14) are convenient for a formulation of a linear conjunction problem corresponding to different mixed conditions at the interface of the materials.

2.1.2 Solution of the Problem for a Periodic Set of Cracks with Contact Zones

Inserting Eq. (2.14) into the boundary conditions (2.3) and (2.4) gives

$$\begin{cases} F^+(x) + \gamma F^-(x) = 0, & x \in M, \\ \mathrm{Im}[F^+(x) + \gamma F^-(x)] = 0, & x \in L, \\ \mathrm{Im}[F^+(x) - F^-(x)] = 0, & x \in L, \end{cases}$$

or

$$\begin{cases} F^+(x) + \gamma F^-(x) = 0, & x \in M, \\ \mathrm{Im}\, F^\pm(x) = 0, & x \in L. \end{cases} \tag{2.15}$$

According to [9] and taking into account the automorphism [8], a canonical solution of the periodic combined Dirichlet-Riemann boundary value problem (2.15) will be searched in the form

$$S(z) = Z(z)e^{i\psi(z)}\Xi^{-\alpha}(z - a), \tag{2.16}$$

where

$$\varXi(\vartheta) = \sin(\pi\vartheta/h),$$

$$Z(z) = \left(\frac{\varXi(z-b)}{\varXi(z-c)}\right)^{1/2-i\varepsilon}$$ - known canonical solution of a periodic Riemann
problem,

α - an integer,

$\varepsilon = (\ln\gamma)/2\pi,$

$\psi(z)$ - solution of a periodic Dirichlet problem,

$$\mathrm{Re}\,\psi^{\pm}(x) = h^{\pm}(x), \quad x \in L, \tag{2.17}$$

$$h^{\pm}(x) = \pi n^{\pm} - \arg Z^{\pm}(x) + \alpha[\arg\varXi(x-a)]^{\pm}, \tag{2.18}$$

which is limited at nodes and at infinity; $n^{\pm}$ are integers.

An automorphic function is an analytical function, the value of which is not changed when its argument is subjected to some sectional-linear transformations. Periodic functions and, particularly, elliptic functions belong to the automorphic functions.

Consider substitutions $\omega_1(z), \ldots, \omega_m(z)$. If one creates a set which contains the given substitutions, the inverse substitutions and all possible combinations constructed from them, then this set forms a group. This group is called to be generated by the substitutions $\omega_1(z), \ldots, \omega_m(z)$, and these substitutions are called the main substitutions of the group. Groups generated by a single substitution are called cyclic.

Rational automorphic functions form a single class of the single-valued functions which are automorphic with respect to finite groups. The points are called equivalent if they are derived from each other through group's substitutions. Automorphic function has the same values at equivalent points. If the region does not contain two different points which are equivalent to each other, but contains all points which are equivalent to any point in the plane with respect to the considered group, then this region is called the fundamental region of the group (the fundamental region of the automorphic function).

A special role in the theory of automorphic functions is played by the so-called main function of the group. This is the automorphic function $H(z) = \sum\limits_{k=1}^{n-1}\dfrac{1}{\omega_k(z) - a}$ (here a is any fixed number), which takes once each of its values in every fundamental region.

In the case of a periodic problem with a period h, the group of transformations is generated by the substitution $\omega_1(z) = z + h$. The fundamental region is a vertical strip of width h, for example, a strip $0 \leqslant \mathrm{Re}\,z \leqslant h$. The main automorphic function is $e^{iz} = e^{-y}e^{ix}$. This function turns to zero at the upper end of the strip, and it has a simple pole at the lower end of the strip. A point at infinity is essentially special for the function e^{iz} and it has a pole of an infinite order as $y \to -\infty$. However, the point at infinity belongs to all fundamental regions. Therefore, the pole's order can be divided among all of them. From a general property of an automorphic function to take each of its values once follows that the pole's order should be considered equal to one at the lower end of each strip. The same is true for the order of zero at the upper end of the strip.

A solution of the problem (2.17) and (2.18) will be constructed [8, 9] taking into account the automorphism properties in the form

$$\psi(z) = \frac{Y(z)}{4\pi i} \int_L \frac{h^+(t) + h^-(t)}{Y^+(t)\Xi(t-z)} dt + \frac{1}{4\pi i} \int_L \left(h^+(t) - h^-(t) \right) \cot \frac{\pi(t-z)}{h} dt,$$

where $Y(z) = \sqrt{\Xi(z-b)\Xi(z-a)}$.

Giving $n^\pm$ and α all possible integer values, the set of canonical solutions are obtained. Only two of them are linearly independent:

$$S_1(z) = \frac{e^{i\varphi(z)}}{\sqrt{\Xi(z-c)\Xi(z-b)}}, \qquad S_2(z) = \frac{i e^{i\varphi(z)}}{\sqrt{\Xi(z-c)\Xi(z-a)}},$$

$$\varphi(z) = -\frac{\varepsilon Y(z)}{2} \int_M \frac{dt}{Y(t)\Xi(t-z)}. \tag{2.19}$$

Then the general solution has the form

$$F(z) = S_1(z)P(e^{iz}) + S_2(z)Q(e^{iz}), \tag{2.20}$$

where $P(e^{iz})$, $Q(e^{iz})$ - are arbitrary polynomials.

To satisfy the conditions at infinity, they must have the following form:

$$P(z) = C_1 \cos[\pi(z - a_*)/h] + C_2 \sin[\pi(z - a_*)/h], \quad a_* = (c+b)/2,$$

$$Q(z) = D_1 \cos[\pi(z - b_*)/h] + D_2 \sin[\pi(z - b_*)/h], \quad b_* = (c+a)/2,$$

where C_1, C_2, D_1, D_2 - are arbitrary real constants.

After insertion of the canonical solutions (2.19) into Eq. (2.20), the general solution (2.20) of the homogeneous combined Dirichlet-Riemann boundary value problem (2.15) takes the form

$$F(z) = \frac{e^{i\varphi(z)}}{\sqrt{\Xi(z-c)}} \left(\frac{P(z)}{\sqrt{\Xi(z-b)}} + i \frac{Q(z)}{\sqrt{\Xi(z-a)}} \right), \tag{2.21}$$

where, after conducting integration in (2.19), the function $\varphi(z)$ takes the following form:

$$\varphi(z) = 2\varepsilon \ln\left(\frac{\sqrt{\Xi(a-b)\Xi(z-c)}}{\sqrt{\Xi(a-c)\Xi(z-b)} + \sqrt{\Xi(b-c)\Xi(z-a)}}\right).$$

Real constants C_1, C_2, D_1, D_2 should be determined from the behavior of the function $F(z)$ at infinity as $z \to \pm i\infty$.

Consider the behavior of the function $\varphi(z)$. At first transform an argument of logarithm which is denoted by

$$R_1(z) = \frac{\sqrt{\Xi(a-b)\Xi(z-c)}}{\sqrt{\Xi(a-c)\Xi(z-b)} + \sqrt{\Xi(b-c)\Xi(z-a)}}.$$

Since the argument of the trigonometric functions is complex and it is necessary to investigate its behavior at infinity for the imaginary axis coordinate (y-axis), the functions dependent on y should be identified separately. Transforming the last expression gives

$$R_1(z) = \frac{\sqrt{\sin\dfrac{\pi(a-b)}{h}}\left[\sin\dfrac{\pi(x-c)}{h} + i\cos\dfrac{\pi(x-c)}{h}\tanh\dfrac{\pi y}{h}\right]}{\left(\begin{array}{l}\sqrt{\sin\dfrac{\pi(a-c)}{h}}\left[\sin\dfrac{\pi(x-b)}{h} + i\cos\dfrac{\pi(x-b)}{h}\tanh\dfrac{\pi y}{h}\right] + \\ + \sqrt{\sin\dfrac{\pi(b-c)}{h}}\left[\sin\dfrac{\pi(x-a)}{h} + i\cos\dfrac{\pi(x-a)}{h}\tanh\dfrac{\pi y}{h}\right]\end{array}\right)}$$

Now tending y to $+\infty$, the following expression can be obtained:

$$R_1(z)\Big|_{z\to +i\infty} = \frac{\sqrt{\sin\dfrac{\pi(a-b)}{h}}\left[\sin\dfrac{\pi(x-c)}{h} + i\cos\dfrac{\pi(x-c)}{h}\right]}{\left(\begin{array}{l}\sqrt{\sin\dfrac{\pi(a-c)}{h}}\left[\sin\dfrac{\pi(x-b)}{h} + i\cos\dfrac{\pi(x-b)}{h}\right] + \\ + \sqrt{\sin\dfrac{\pi(b-c)}{h}}\left[\sin\dfrac{\pi(x-a)}{h} + i\cos\dfrac{\pi(x-a)}{h}\right]\end{array}\right)}$$

and, simplifying,

$$R_1(z)\Big|_{z\to +i\infty} = \frac{\sqrt{\Xi(a-b)}\,e^{-i\frac{\pi(x-c)}{2h}}}{\sqrt{\Xi(a-c)}\,e^{-i\frac{\pi(x-b)}{2h}} + \sqrt{\Xi(b-c)}\,e^{-i\frac{\pi(x-a)}{2h}}}.$$

The variable x is mutually reduced in the numerator and denominator, therefore a constant is determined at infinity

$$R_1(z)\bigg|_{z\to+i\infty} = \frac{\sqrt{\Xi(a-b)}e^{i\frac{\pi c}{2h}}}{\sqrt{\Xi(a-c)}e^{i\frac{\pi b}{2h}} + \sqrt{\Xi(b-c)}e^{i\frac{\pi a}{2h}}}.$$

Similarly for $y \to -\infty$ follows that

$$R_1(z)\bigg|_{z\to i\infty} = \frac{\sqrt{\Xi(a-b)}e^{-i\frac{\pi c}{2h}}}{\sqrt{\Xi(a-c)}e^{-i\frac{\pi b}{2h}} + \sqrt{\Xi(b-c)}e^{-i\frac{\pi a}{2h}}}.$$

Therefore

$$\varphi(z)\bigg|_{y\to+\infty} = \exp\left[i\cdot 2\varepsilon \ln \frac{\sqrt{\Xi(a-b)}e^{i\frac{\pi c}{2h}}}{\sqrt{\Xi(a-c)}e^{i\frac{\pi b}{2h}} + \sqrt{\Xi(b-c)}e^{i\frac{\pi a}{2h}}}\right]$$

or

$$\varphi(z)\bigg|_{y\to+\infty} = \exp\left[-i\cdot 2\varepsilon \ln \frac{\sqrt{\Xi(a-c)}e^{i\frac{\pi(b-c)}{2h}} + \sqrt{\Xi(b-c)}e^{i\frac{\pi(a-c)}{2h}}}{\sqrt{\Xi(a-b)}}\right].$$

It is necessary to identify separately the real and imaginary parts of the value $\varphi(z)$ at infinity to form a system of linear algebraic equations. Some transformations will be done for this purpose.

Proceed to the trigonometric form:

$$\varphi(z)\bigg|_{y\to+\infty} = e^{-i\cdot 2\varepsilon \ln \bar{\omega}},$$

$$\bar{\omega} = \frac{\left(\sqrt{\sin\frac{\pi(a-c)}{h}}\left[\cos\frac{\pi(b-c)}{2h} + i\sin\frac{\pi(b-c)}{2h}\right] + \right.}{\left. +\sqrt{\sin\frac{\pi(b-c)}{h}}\left[\cos\frac{\pi(a-c)}{2h} + i\sin\frac{\pi(a-c)}{2h}\right]\right)}{\sqrt{\sin\frac{\pi(a-b)}{h}}}.$$

Using the logarithm's properties of a complex number $\ln z = \ln|z| + i\,\mathrm{Arg}\,z$, after the corresponding transformations and using the elementary theorems of the complex variable theory, it follows that

$$\varphi(z)\Big|_{y\to+\infty} =$$

$$= \exp\left[\begin{array}{c} -i\cdot\varepsilon\ln\dfrac{\varXi\left(\frac{a+b}{2}-c\right)+\sqrt{\varXi(a-c)\varXi(b-c)}}{\varXi\left(\frac{a-b}{2}\right)}+ \\[2em] +2\varepsilon\arctan\dfrac{\sqrt{\varXi(a-c)}\,\varXi\left(\frac{b-c}{2}\right)+\sqrt{\varXi(b-c)}\,\varXi\left(\frac{a-c}{2}\right)}{\sqrt{\sin\dfrac{\pi(a-c)}{h}}\cos\dfrac{\pi(b-c)}{2h}+\sqrt{\sin\dfrac{\pi(b-c)}{h}}\cos\dfrac{\pi(a-c)}{2h}} \end{array}\right]$$

or

$$\varphi(z)\Big|_{y\to+\infty} =$$

$$= \exp\left[2\varepsilon\arctan\dfrac{\sqrt{\sin\dfrac{\pi(a-c)}{h}}\sin\dfrac{\pi(b-c)}{2h}+\sqrt{\sin\dfrac{\pi(b-c)}{h}}\sin\dfrac{\pi(a-c)}{2h}}{\sqrt{\sin\dfrac{\pi(a-c)}{h}}\cos\dfrac{\pi(b-c)}{2h}+\sqrt{\sin\dfrac{\pi(b-c)}{h}}\cos\dfrac{\pi(a-c)}{2h}}\right]\times$$

$$\times\left[\begin{array}{c}\cos\left\{\varepsilon\ln\dfrac{\sin\dfrac{\pi(a+b-2c)}{2h}+\sqrt{\sin\dfrac{\pi(a-c)}{h}\sin\dfrac{\pi(b-c)}{h}}}{\sin\dfrac{\pi(a-b)}{2h}}\right\}- \\[3em] -i\sin\left\{\varepsilon\ln\dfrac{\sin\dfrac{\pi(a+b-2c)}{2h}+\sqrt{\sin\dfrac{\pi(a-c)}{h}\sin\dfrac{\pi(b-c)}{h}}}{\sin\dfrac{\pi(a-b)}{2h}}\right\}\end{array}\right].$$

Introducing new real constants by the formulae

$$\chi = 2\varepsilon\arctan\dfrac{\sqrt{\sin\dfrac{\pi(a-c)}{h}}\sin\dfrac{\pi(b-c)}{2h}+\sqrt{\sin\dfrac{\pi(b-c)}{2h}}\sin\dfrac{\pi(a-c)}{2h}}{\sqrt{\sin\dfrac{\pi(a-c)}{h}}\cos\dfrac{\pi(b-c)}{2h}+\sqrt{\sin\dfrac{\pi(b-c)}{h}}\cos\dfrac{\pi(a-c)}{2h}},$$

$$\zeta = \varepsilon\ln\dfrac{\sin\dfrac{\pi(a+b-2c)}{2h}+\sqrt{\sin\dfrac{\pi(a-c)}{h}\sin\dfrac{\pi(b-c)}{h}}}{\sin\dfrac{\pi(a-b)}{2h}},$$

$$\tag{2.22}$$

we finally obtain that

$$\varphi(z)\Big|_{y\to+\infty} = e^{\chi}(\cos\zeta - i\sin\zeta). \tag{2.23}$$

The similar analysis for $y \to -\infty$ leads to the expression

$$\varphi(z)\Big|_{y \to -\infty} = e^{-\chi}(\cos \zeta - i \sin \zeta). \tag{2.24}$$

Now investigate the behavior at infinity of another part of the function $F(z)$, namely, the expression

$$R_2(z) = \frac{P(z)}{\sqrt{\Xi(z - c)\Xi(z - b)}}.$$

Inserting the expression for $P(z)$ gives

$$R_2(z) = \frac{C_1 \cos \dfrac{\pi(z - \frac{c+b}{2})}{h} + C_2 \sin \dfrac{\pi(z - \frac{c+b}{2})}{h}}{\sqrt{\sin \dfrac{\pi(z - c)}{h} \sin \dfrac{\pi(z - b)}{h}}}.$$

Transforming the last expression to get rid of the complex argument of trigonometric functions, brings

$$R_2(z) =$$

$$= \frac{\left(C_1 \left[\cos \dfrac{\pi(x - \frac{c+b}{2})}{h} - i \sin \dfrac{\pi(x - \frac{c+b}{2})}{h} \tanh \dfrac{\pi y}{h} \right] + \atop +C_2 \left[\sin \dfrac{\pi(x - \frac{c+b}{2})}{h} + i \cos \dfrac{\pi(x - \frac{c+b}{2})}{h} \tanh \dfrac{\pi y}{h} \right] \right)}{\sqrt{\left[\sin \dfrac{\pi(x - c)}{h} + i \cos \dfrac{\pi(x - c)}{h} \tanh \dfrac{\pi y}{h} \right] \left[\sin \dfrac{\pi(x - b)}{h} + i \cos \dfrac{\pi(x - b)}{h} \tanh \dfrac{\pi y}{h} \right]}}.$$

Tending y to $+\infty$, the following formula is obtained:

$$R_2(z)\Big|_{z \to +i\infty} =$$

$$= \frac{C_1 \left[\cos \dfrac{\pi(x - \frac{c+b}{2})}{h} - i \sin \dfrac{\pi(x - \frac{c+b}{2})}{h} \right] + C_2 \left[\sin \dfrac{\pi(x - \frac{c+b}{2})}{h} + i \cos \dfrac{\pi(x - \frac{c+b}{2})}{h} \right]}{\sqrt{\left[\sin \dfrac{\pi(x - c)}{h} + i \cos \dfrac{\pi(x - c)}{h} \right] \left[\sin \dfrac{\pi(x - b)}{h} + i \cos \dfrac{\pi(x - b)}{h} \right]}}$$

or, after transformations,

$$R_2(z)\Big|_{z \to +i\infty} = \frac{(C_1 + iC_2)e^{-i\frac{\pi(x - \frac{c+b}{2})}{h}}}{ie^{-i\frac{\pi(2x - c - b)}{2h}}}.$$

It is important that the variable x is mutually reduced in the numerator and denominator, so

$$R_2(z)\Big|_{z\to+i\infty} = C_2 - iC_1. \tag{2.25}$$

Similarly, for y tending to $-\infty$ follows that

$$R_2(z)\Big|_{z\to-i\infty} = C_2 + iC_1. \tag{2.26}$$

Investigate the behavior of the rest of the function $F(z)$ as $y \to +\infty$ and as $y \to -\infty$, namely, of the expression

$$R_3(z) = i\frac{Q(z)}{\sqrt{\Xi(z-c)\Xi(z-a)}}.$$

Substitution of the expression for $Q(z)$ gives

$$R_3(z) = i\,\frac{D_1 \cos\dfrac{\pi(z-\frac{c+a}{2})}{h} + D_2 \sin\dfrac{\pi(z-\frac{c+a}{2})}{h}}{\sqrt{\sin\dfrac{\pi(z-c)}{h}\sin\dfrac{\pi(z-a)}{h}}}.$$

Transforming the above expression by expanding the complex argument leads to the formula

$$R_3(z) = $$

$$= \frac{\left(D_1\left[\cos\dfrac{\pi(x-\frac{c+a}{2})}{h} - i\sin\dfrac{\pi(x-\frac{c+a}{2})}{h}\tanh\dfrac{\pi y}{h}\right] + \right.}{\left.\quad + D_2\left[\sin\dfrac{\pi(x-\frac{c+a}{2})}{h} + i\cos\dfrac{\pi(x-\frac{c+a}{2})}{h}\tanh\dfrac{\pi y}{h}\right]\right)}$$

$$\overline{\sqrt{\left[\sin\dfrac{\pi(x-c)}{h} + i\cos\dfrac{\pi(x-c)}{h}\tanh\dfrac{\pi y}{h}\right]\left[\sin\dfrac{\pi(x-a)}{h} + i\cos\dfrac{\pi(x-a)}{h}\tanh\dfrac{\pi y}{h}\right]}}.$$

And as $y \to +\infty$

$$R_3(z)\Big|_{z\to+i\infty} =$$

$$= \frac{D_1\left[\cos\dfrac{\pi(x-\frac{c+a}{2})}{h} - i\sin\dfrac{\pi(x-\frac{c+a}{2})}{h}\right] + D_2\left[\sin\dfrac{\pi(x-\frac{c+a}{2})}{h} + i\cos\dfrac{\pi(x-\frac{c+a}{2})}{h}\right]}{\sqrt{\left[\sin\dfrac{\pi(x-c)}{h} + i\cos\dfrac{\pi(x-c)}{h}\right]\left[\sin\dfrac{\pi(x-a)}{h} + i\cos\dfrac{\pi(x-a)}{h}\right]}}$$

or, after transformations,

$$R_3(z)\bigg|_{z\to+i\infty} = \frac{(D_1 + i D_2)e^{-i\frac{\pi(x-\frac{c+a}{2})}{h}}}{ie^{-i\frac{\pi(2x-c-a)}{2h}}}.$$

As before, the variable x is mutually eliminated in the numerator and denominator, so a constant value is obtained

$$R_3(z)\bigg|_{z\to+i\infty} = D_1 + i D_2. \tag{2.27}$$

Similarly, for y tending to $-\infty$ follows that

$$R_3(z)\bigg|_{z\to-i\infty} = -D_1 + i D_2. \tag{2.28}$$

The expression of the function $F(z)$ at infinity is obtained, collecting the expressions (2.23)–(2.28) into (2.21):

$$F(z)\bigg|_{z\to+i\infty} = e^{\chi}(\cos\zeta - i\sin\zeta)[C_2 - iC_1 + D_1 + iD_2], \tag{2.29}$$

$$F(z)\bigg|_{z\to-i\infty} = e^{-\chi}(\cos\zeta - i\sin\zeta)[C_2 + iC_1 - D_1 + iD_2]. \tag{2.30}$$

Equating expression (2.13) with Eqs. (2.29) and (2.30), a system of linear algebraic equations is obtained:

$$\begin{cases} e^{\chi}(\cos\zeta - i\sin\zeta)\big[C_2 - iC_1 + D_1 + iD_2\big] = \dfrac{\sigma - i\tau}{g(1+\gamma)}, \\[2mm] e^{-\chi}(\cos\zeta - i\sin\zeta)\big[C_2 + iC_1 - D_1 + iD_2\big] = \dfrac{\sigma - i\tau}{g(1+\gamma)}. \end{cases}$$

Solution of this system determines the arbitrary real constants C_1, C_2, D_1, D_2:

$$C_1 = \frac{\sinh\chi(-\tau\cos\zeta + \sigma\sin\zeta)}{g(1+\gamma)}, \quad C_2 = \frac{\cosh\chi(\sigma\cos\zeta + \tau\sin\zeta)}{g(1+\gamma)},$$

$$D_1 = \frac{-\sinh\chi(\sigma\cos\zeta + \tau\sin\zeta)}{g(1+\gamma)}, \quad D_2 = \frac{\cosh\chi(-\tau\cos\zeta + \sigma\sin\zeta)}{g(1+\gamma)}. \tag{2.31}$$

2.1.3 Derivation of the Classical "Oscillating" Solution as a Particular Case of the "Contact" Model Solution

The solution within the contact model can be analytically reduced to the oscillating one on assumption that the contact zone length tends to zero, namely:

$$
\left. F(z) \right|_{b \to a} =
$$

$$
= e^{i \cdot 2\varepsilon \ln \frac{\sqrt{\Xi(a-b)\Xi(z-c)}}{\sqrt{\Xi(a-c)\Xi(z-a)}}} \left. \frac{[C_1 + i D_1] \cos \dfrac{\pi(z - \frac{c+a}{2})}{h} + [C_2 + i D_2] \sin \dfrac{\pi(z - \frac{c+a}{2})}{h}}{\sqrt{\Xi(z-c)\Xi(z-a)}} \right|_{b \to a},
$$

or, using formula (2.31),

$$
\left. F(z) \right|_{b \to a} = \frac{\sigma - i\tau}{g(1+\gamma)} e^{i\varepsilon \ln \frac{\Xi(a-b)\Xi(z-c)}{\Xi(a-c)\Xi(z-a)}} \times
$$

$$
\times \left. \frac{[-i \cos \zeta + \sin \zeta] \sinh \chi \cos \dfrac{\pi(z - \frac{c+a}{2})}{h} + [\cos \zeta + i \sin \zeta] \cosh \chi \sin \dfrac{\pi(z - \frac{c+a}{2})}{h}}{\sqrt{\Xi(z-c)\Xi(z-a)}} \right|_{b \to a}.
$$

After elementary transformations

$$
\left. F(z) \right|_{b \to a} =
$$

$$
= e^{i\varepsilon \ln \frac{\Xi(a-b)\Xi(z-c)}{\Xi(a-c)\Xi(z-a)}} e^{i\zeta} \left. \frac{-i \sinh \chi \cos \dfrac{\pi(z - \frac{c+a}{2})}{h} + \cosh \chi \sin \dfrac{\pi(z - \frac{c+a}{2})}{h}}{\sqrt{\Xi(z-c)\Xi(z-a)}} \frac{\sigma - i\tau}{g(1+\gamma)} \right|_{b \to a}.
$$

Then, using formula (2.22) and taking into account the behavior of the functions ζ and χ as $b \to a$, it is derived

$$
\left. F(z) \right|_{b \to a} = \frac{-i \sin \dfrac{i\pi\varepsilon(a-c)}{h} \cos \dfrac{\pi(z - \frac{c+a}{2})}{h} + \cos \dfrac{i\pi\varepsilon(a-c)}{h} \sin \dfrac{\pi(z - \frac{c+a}{2})}{h}}{\sqrt{\Xi(z-c)\Xi(z-a)}}
$$

$$
\times \left[\frac{\Xi(z-c)}{\Xi(z-a)} \right]^{i\varepsilon} \frac{\sigma - i\tau}{g(1+\gamma)}
$$

and, finally,

$$
\left. F(z) \right|_{b \to a} = \frac{\Xi \left(z - \frac{c+a}{2} - i\varepsilon(a-c) \right)}{\sqrt{\Xi(z-c)\Xi(z-a)}} \left[\frac{\Xi(z-c)}{\Xi(z-a)} \right]^{i\varepsilon} \frac{\sigma - i\tau}{g(1+\gamma)}. \tag{2.32}
$$

The obtained expression for the function $F(z)$ coincides with the known formula [10] for the classical model.

2.1.4 Determination of the Contact Zone Length and the Stress Intensity Factors

Using formula (2.14), the expressions for the stresses and the displacement jump at different parts of the material interface are:

$$\sigma_{22}^{(1)}(x, 0) = \frac{2ge^{\pi\varepsilon}}{\sqrt{\Xi(x-c)}}\left(\frac{P(x)\cosh(\tilde{\varphi}(x) - \pi\varepsilon)}{\sqrt{\Xi(x-b)}} + \frac{Q(x)\sinh(\tilde{\varphi}(x) - \pi\varepsilon)}{\sqrt{\Xi(a-x)}}\right), \quad x \in L,$$

(2.33)

$$\sigma_{22}^{(1)}(x, 0) - i\sigma_{12}^{(1)}(x, 0) = \frac{g(1+\gamma)e^{i\varphi(x)}}{\sqrt{\Xi(x-c)}}\left(\frac{P(x)}{\sqrt{\Xi(x-b)}} + i\frac{Q(x)}{\sqrt{\Xi(x-a)}}\right), \quad x \in U,$$

(2.34)

$$\langle u_2'(x)\rangle = -\frac{2\cosh(\pi\varepsilon)}{\sqrt{\Xi(x-c)}}\left(\frac{P(x)}{\sqrt{\Xi(b-x)}}\cos(\varphi^*(x)) - \frac{Q(x)}{\sqrt{\Xi(a-x)}}\sin(\varphi^*(x))\right), \quad x \in M,$$

(2.35)

where

$$\tilde{\varphi}(x) = 2\varepsilon\arctan\sqrt{\frac{\Xi(b-c)\Xi(a-x)}{\Xi(a-c)\Xi(x-b)}}, \quad x \in L,$$

$$\varphi^*(x) = 2\varepsilon\ln\left(\frac{\sqrt{\Xi(a-b)\Xi(x-c)}}{\sqrt{\Xi(a-c)\Xi(b-x)} + \sqrt{\Xi(b-c)\Xi(a-x)}}\right), \quad x \in M.$$

Determination of the contact zone length. Expressions (2.33)–(2.35) can be used for any position of the point b, but in oder that the obtained solution will be physically sound, additional conditions must be satisfied

$$\langle u_2'(b)\rangle = 0; \quad \sigma_{22}^{(1)}(x, 0) \le 0, \quad x \in L; \quad \langle u_2(x)\rangle \ge 0, \quad x \in M. \tag{2.36}$$

The first equation in (2.36) shows that the crack faces close smoothly. The second and third inequalities indicate that the normal stresses are compressive in contact zones and the crack faces do not overlap each other.

Expanding Eqs. (2.33) and (2.35) into Taylor series in the vicinity of the point b and using condition (2.36₁), the transcendental equation to determine the relative contact zone length is derived

$$P(b) = 0. \tag{2.37}$$

Correct selection of its root is insured by satisfaction of conditions (2.36$_2$) and (2.36$_3$). Eq. (2.37) can be written as

$$\tan\frac{\pi(b-c)}{2h} = -\frac{-\tau\cos\zeta + \sigma\sin\zeta}{\sigma\cos\zeta + \tau\sin\zeta}\tanh\chi.$$

SIFs at the crack tip a are introduced as following:

$$K_1 - iK_2 = \lim_{x\to a}\left(\sigma_{22}(x,0) - i\sigma_{12}(x,0)\right)\sqrt{\Xi(x-a)}. \tag{2.38}$$

Consequently, using Eq. (2.34), the following expressions for the SIFs are obtained:

$$K_1 = 0,$$

$$
K_2 = -\frac{g(1+\gamma)}{\sqrt{\Xi(a-c)}}\left(D_1\cos\frac{\pi(a-c)}{2h} + D_2\sin\frac{\pi(a-c)}{2h}\right) =
$$
$$
= \frac{(\sigma\cos\zeta + \tau\sin\zeta)\sinh\chi\cos\frac{\pi(a-c)}{2h} + (\tau\cos\zeta - \sigma\sin\zeta)\cosh\chi\sin\frac{\pi(a-c)}{2h}}{\sqrt{\Xi(a-c)}}.
$$
$$\tag{2.39}$$

It should be noted that in the framework of the classical model, the expressions for the SIFs are obtained in the form, which is identical to [10]:

$$K_1^{osc} - iK_2^{osc} = [\Xi(a-c)]^{-1/2+i\varepsilon}\Xi[(a-c)(1/2-i\varepsilon)]\operatorname{sech}(\pi\varepsilon)(\sigma - i\tau).$$

In case of a homogeneous material the last expression can be simplified to

$$K_1 - iK_2 = \frac{\Xi((a-c)/2)}{\sqrt{\Xi(a-c)}}(\sigma - i\tau).$$

2.1.5 Numerical Results

To investigate an interaction in a periodic set of cracks it is enough to consider the influence of the crack length (l) on the relative contact zone length $\lambda = (a-b)/l$ and on the SIFs for various values of a loading parameter β and an elastic parameter of a bimaterial ε. From the definition of these parameters (β is an angle between the y-axis and the resulting loading $p = \sqrt{\sigma^2 + \tau^2}$; ε characterizes the relative stiffness of materials) follows that they vary in the range: $-\pi/2 \le \beta \le \pi/2$, $-(\ln 3)/2\pi \le \varepsilon \le (\ln 3)/2\pi$. Positive (negative) values of the parameter ε mean that the "lower" ("upper") material is stiffer than the "upper" ("lower") one. max/min values of ε

Fig. 2.2 Dependence of the relative contact zone length λ on the distance between the cracks under tensile loading

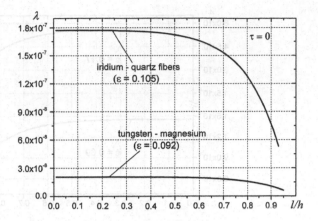

Fig. 2.3 Dependence of the relative contact zone length λ on the distance between the cracks under combined loading

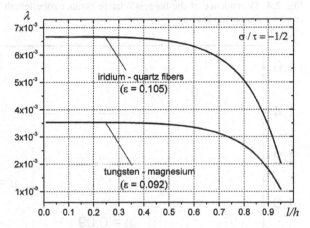

correspond to the case when one of materials is absolutely rigid (Young's modulus tends to infinity) and Poisson's ratio of the second material is equal to 0.

Figures 2.2 and 2.3 show the dependence of the relative contact zone length on the ratio of crack length to the period for combinations of materials tungsten - magnesium and iridium - quartz fibers. It can be seen that for the combination of materials with greater relative stiffness, characterized by the parameter ε, the contact zone length is larger. The contact zone increases as well in the presence of the shear loading (acting in the negative direction in case when upper material is less rigid).

The dependence of the relative contact zone length λ on the distance between the cracks for the tensile loading ($\tau = 0$) and different values of ε is plotted in Fig. 2.4. In one case, the value of ε is the biggest possible: $\varepsilon = 0.17$. The relative contact zone length decreases with increasing ratio of the crack length to the period. The relative contact zone length increases with increasing relative stiffness parameter ε.

For the tensile loading ($\tau = 0$) the dependence of the dimensionless SIF $K^* = -K_2/\sqrt{(\sigma^2 + \tau^2)l}$ on the distance between the cracks is plotted in Fig. 2.5. As it is

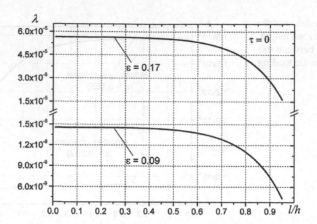

Fig. 2.4 Dependence of the biggest relative contact zone length λ on the distance between the cracks under tensile loading

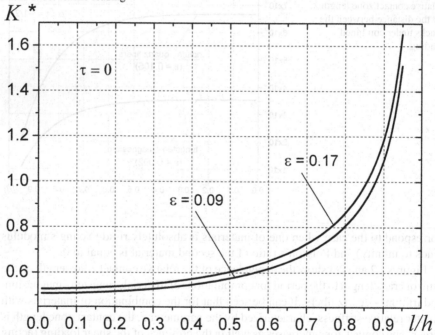

Fig. 2.5 Dependence of the dimensionless SIF K^* on the distance between the cracks under tensile loading

expected, the SIFs increase with decreasing the distance between the cracks and by increasing the relative stiffness parameter ε.

The results for the relative contact zone length under the combined loading are presented in Fig. 2.6. Due to the fact that the upper material is less rigid than the lower one, an increase of the shear loading leads to the contact zone increase. This is seen by comparing the results presented in Figs. 2.4 and 2.6. For the illustrative purposes, an axis break is used along the ordinate axis in Figs. 2.4 and 2.6 because contact zone length can differ some orders of magnitude for various bimaterials.

Table 2.1 shows the results of the comparative analysis of the relative contact zone length when the crack length is much smaller than the distance between the cracks (λ^0), and the known results for a single crack [11] (λ^*). For all values of the parameter ε the results for λ^0 and λ^* are almost identical which indirectly confirms the correctness of the solution for the periodic set of cracks.

A comparison of the normal stresses along the interface of materials ahead of the cracks is presented in Fig. 2.7. Normal stress fields are calculated within the framework of the oscillating and the contact models under the combined loading for a pair of tungsten-magnesium. The ratio of the tensile to shear loading is -2 ($\sigma = 1$ MPa, $\tau = -2$ MPa), the scale for stresses in Fig. 2.7 is MPa. The crack length is half of a period. The relative contact zone length under such conditions is equal to 3.448×10^{-3}, $x_1' = (x_1 - a)/l$. The results show that the stresses start to coincide from a distance of 5 contact zone lengths.

A comparison of the tangential stresses along the interface of materials ahead of the cracks is presented in Fig. 2.8. Tangential stress fields are calculated within the framework of the oscillating and the contact models under the combined loading. A

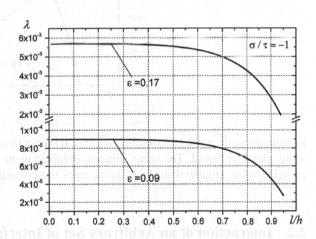

Fig. 2.6 Dependence of the relative contact zone length λ for the biggest $\varepsilon = 0.17$ on the distance between the cracks under combined loading

Table 2.1 Contact zone length for a single crack (λ^*) and for a periodic set of cracks (λ^0) at $l/h = 1/20$ under tensile loading

ε	0.09	0.11	0.13	0.15	0.17
λ^*	1.45481×10^{-8}	3.51059×10^{-7}	3.19656×10^{-6}	1.62284×10^{-5}	5.64793×10^{-5}
λ^0	1.45480×10^{-8}	3.51058×10^{-7}	3.19655×10^{-6}	1.62283×10^{-5}	5.64792×10^{-5}

Fig. 2.7 Comparison of the normal stress fields (oscillatory vs. contact approaches) near the right crack tip under combined loading

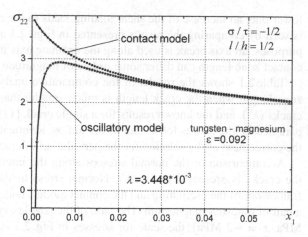

Fig. 2.8 Comparison of the tangential stress fields (oscillatory vs. contact approaches) near the right crack tip under combined loading

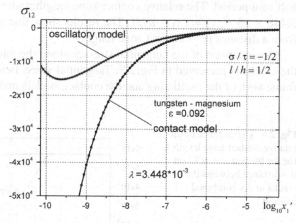

logarithmic scale is used for the abscissa axis. It is easy to see the oscillating behavior of the classical model. The shear stresses, obtained using different models, start to coincide from a smaller distance from the crack tips compared to the normal stresses.

2.2 Interaction of an Arbitrary Set of Interface Cracks

2.2.1 Statement and Solution of the Dirichlet-Riemann BVP for a Set of Interface Cracks with Contact Zones

Statement of the problem. Consider a finite set of the cracks located at the interface of two dissimilar half-planes or half-spaces. The materials are assumed to be isotropic with Young's moduli E_k and Poisson's ratios v_k, where $k = 1$ is related to the upper

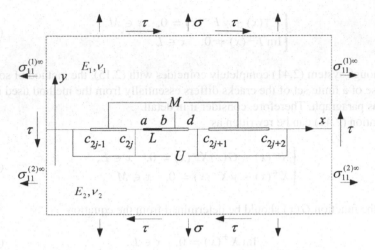

Fig. 2.9 Finite set of the interface cracks between two isotropic materials

material and $k = 2$ to the lower one. The plane is subjected to an uniform tensile (σ) and shear (τ) loading. In determination of the SIFs at some crack tip only the contact zone which is located near this tip can be taken into account.

The following denotation is introduced further: the point between the contact zone and the open crack faces is denoted by b, the point between the bonded interface and the contact zone - by a and the point between the bonded interface and the open crack faces - by d. Other crack tips are denoted by $c_j (j = 1, ..., J - 1)$. Other cracks may have contact zones as well. The union of the open crack faces is denoted by M, the set of the contact zones - by L and the bonded parts of the interface - by U. Assuming that the contact zones are smooth and the open parts of the cracks are load-free, the continuity and boundary conditions in Cartesian coordinates xy, depicted in Fig. 2.9, can be written as (2.1)–(2.4) (cf. previous section).

After transformations, similar to (2.5)–(2.9), the fields of stresses and displacements (2.10)–(2.12) are expressed via one function $F(z)$ which is analytical in the entire plane except for $L \cup M$ and infinity where the behavior of the function is represented by the formula (2.13).

Considering the fields of stresses and displacements defined by formulae (2.10)–(2.12) at the interface $y = 0$, the following expressions can be written

$$\sigma_{22}^{(1)}(x, 0) - i\sigma_{12}^{(1)}(x, 0) = g[F^+(x) + \gamma F^-(x)],$$
$$\langle u_1'(x) \rangle + i\langle u_2'(x) \rangle = F^+(x) - F^-(x). \tag{2.40}$$

Inserting relations (2.40) into boundary conditions (2.3) and (2.4) gives

$$\begin{cases} F^+(x) + \gamma F^-(x) = 0, & x \in M, \\ \operatorname{Im} F^\pm(x) = 0, & x \in L. \end{cases} \tag{2.41}$$

Although system (2.41) completely coincides with (2.15), the method of solving it in case of a finite set of the cracks differs essentially from the method used in the previous paragraph. Therefore, consider it in detail.

Equation (2.41) can be rewritten as

$$\begin{cases} X^+(x) - G(x)X^-(x) = 0, & x \in L, \\ X^+(x) + \gamma X^-(x) = 0, & x \in M, \end{cases} \tag{2.42}$$

where the function $G(x)$ should be determined from the equation

$$\operatorname{Im} X^\pm(x) = 0, \quad x \in L. \tag{2.43}$$

Solution to (2.42) can be written, according to [8], in the form

$$X(x) = \exp\left(\frac{1}{2\pi i} \int\limits_L \frac{\ln [G(x)]dx}{x - z} + \frac{\ln (-\gamma)}{2\pi i} \int\limits_M \frac{dx}{x - z} \right),$$

or, after some transformations,

$$X(z) = e^{\psi(z)}, \quad \psi(z) = \left(\frac{1}{2} - i\varepsilon \right) I(z) - i\Gamma_1(z) + \Gamma_2(z), \tag{2.44}$$

where

$$I(z) = \int\limits_M \frac{dx}{x-z},$$

$$\Gamma_1(z) = \frac{1}{2\pi} \int\limits_L \frac{\ln |G(x)|}{x-z} dx,$$

$$\Gamma_2(z) = \frac{1}{2\pi} \int\limits_L \frac{\arg[G(x)]}{x-z} dx.$$

The boundary values of the function $X(z)$ on $x \in L$ can be found applying Sokhotsky-Plemelj formulae [8] to the integrals $\Gamma_1(z)$ and $\Gamma_2(z)$. The corresponding formulae can be represented as

$$X^\pm(x) = e^{\psi^\pm(x)},$$

$$\psi^\pm(x) = \left(\frac{1}{2} - i\varepsilon \right) I(z) - i\Gamma_1(z) + \Gamma_2(z) \pm \frac{1}{2}(\ln |G(x)| + i \arg[G(x)]), \quad x \in L. \tag{2.45}$$

Inserting (2.45) into (2.43), the following expression can be written

$$\text{Im } \psi^{\pm}(x) = \pi n^{\pm}, \quad x \in L,$$

or

$$-\varepsilon I(x) - \Gamma_1(x) \pm \arg[G(x)] = \pi n^{\pm}, \quad x \in L,$$

where $n^{\pm}$ are arbitrary integers.

The set of two equations written above can be rearranged in the form

$$\arg[G(x)] = \pi(n^+ - n^-), \quad 2\Gamma_1(x) = g(x), \quad g(x) = \pi(n^+ + n^-) - 2\varepsilon I(x), \quad x \in L. \tag{2.46}$$

Equation (2.46$_2$) is a singular integral equation, a solution of which according to [8] can be presented as

$$\Gamma_1(z) = \frac{Z(z)}{2\pi i} \int_L \frac{g(x)dx}{Z^+(x)(z-x)}, \quad Z(z) = \sqrt{(z-a)(z-b)}.$$

Inserting $g(x)$, defined by formula (2.46$_3$), into the previous integral and evaluating it, the following formula is obtained:

$$\Gamma_1(z) = \frac{\pi(n^+ + n^-)}{2} - \varphi(z) - \varepsilon I(z), \tag{2.47}$$

where

$$I(z) = \text{sgn}(d-b) \ln\left(\frac{z-d}{z-b}\right) + \sum_{j=1}^{J-1} \ln\left(\frac{z-c_{2j}}{z-c_{2j-1}}\right), \quad \chi(z) = \frac{z-b}{z-a},$$

$$\varphi(z) = \text{sgn}(d-b)\left(\pi i \varepsilon - 2\varepsilon \operatorname{arctanh}\sqrt{\frac{\chi(d)}{\chi(z)}}\right) -$$

$$- 2\varepsilon \sum_{j=1}^{2J-2} (-1)^j \, \text{sgn}(c_j - b) \operatorname{arctanh}\sqrt{\frac{\chi(c_j)}{\chi(z)}}.$$

Using (2.44$_2$) and (2.47), the function $\psi(z)$ takes the form

$$\psi(z) = \frac{1}{2} I(z) + i\varphi(z) - \frac{i\pi(n^+ + n^-)}{2} + \text{sgn}(b-a)\frac{n^+ - n^-}{2} \ln\left(\frac{z-b}{z-a}\right).$$

Therefore

$$X(z) = \exp\left(i\varphi(z) - \frac{i\pi(n^+ + n^-)}{2}\right) \prod_{j=1}^{J}\left(\frac{z - c_{2j}}{z - c_{2j-1}}\right)^{1/2}\left(\frac{z - b}{z - a}\right)^{\alpha_1}\left(\frac{z - d}{z - b}\right)^{\alpha_2},$$

(2.48)

here

$$\alpha_1 = \text{sgn}(b - a)\frac{n^+ - n^-}{2}, \quad \alpha_2 = \frac{\text{sgn}(d - b)}{2}.$$

It can be easily shown that formula

$$F(z) = \frac{e^{i\varphi(z)}}{f(z)}\left(\frac{P(z)}{\sqrt{z - b}} + i\frac{Q(z)}{\sqrt{z - a}}\right)$$

(2.49)

includes all solutions presented in (2.48).

Here $f(z) = \sqrt{z - d}\prod_{j=1}^{2J-2}(z - c_j)^{1/2}$, $P(z)$ and $Q(z)$ are polynomials with arbitrary real coefficients.

In order for the function (2.49) to be the required solution of the boundary value problem (2.41) and taking into account the condition (2.43), the polynomials $P(z)$ and $Q(z)$ must have the form

$$P(z) = \sum_{j=0}^{J} C_j z^j, \quad Q(z) = \sum_{j=0}^{J} D_j z^j.$$

(2.50)

The coefficients $C_J, D_J, C_{J-1}, D_{J-1}$ can be found from the condition at infinity (2.13) and presented as

$$C_J = \tilde{p}\cos(\alpha_0 + \beta), \quad C_{J-1} = -\tilde{p}\left(\alpha_1\sin(\alpha_0 + \beta) + \upsilon\cos(\alpha_0 + \beta)\right),$$
$$D_J = -\tilde{p}\sin(\alpha_0 + \beta), \quad D_{J-1} = -\tilde{p}\left(\alpha_1\cos(\alpha_0 + \beta) - (\upsilon + \eta)\sin(\alpha_0 + \beta)\right),$$

(2.51)

where

$$\upsilon = \frac{1}{2}\left(b + d + \sum_{j=1}^{2J-2} c_j\right), \quad \eta = \frac{a - b}{2},$$

$$\alpha_0 = 2\varepsilon\,\text{sgn}(d - b)\,\text{arctanh}\sqrt{\chi(d)} + 2\varepsilon\sum_{j=1}^{2J-2}(-1)^j\,\text{sgn}(c_j - b)\,\text{arctanh}\sqrt{\chi(c_j)},$$

$$\alpha_1 = \varepsilon\sqrt{(d - a)(d - b)} - \varepsilon\sum_{j=1}^{2J-2}(-1)^j\,\text{sgn}(c_j - b)\sqrt{(c_j - a)(c_j - b)},$$

$$\tilde{p} = \frac{p}{1+\gamma}, \quad p = \sqrt{\sigma^2 + \tau^2}.$$

The remaining coefficients can be found from a single-valuedness condition of displacements, which, based on (2.40$_2$), can be presented in the following form

$$\int\limits_{L+M} \left[F^+(x) - F^-(x) \right] dx = 0.$$

Using Sokhotsky-Plemelj formulae [8], the following equation can be written

$$F(z) = \frac{1}{2\pi i} \int\limits_{L+M} \frac{F^+(x) - F^-(x)}{x - z} dx + const,$$

and, expanding the above formula into power series at infinity, it follows that

$$F(z) = const - \left\{ \frac{1}{2\pi i} \int\limits_{L+M} \left[F^+(x) - F^-(x) \right] dx \right\} \frac{1}{z} + O\left(\frac{1}{z^2} \right), \quad z \to \infty.$$

Following the comparison of the last formula with the single-valuedness condition of the displacements, the coefficient in the expansion of the function $F(z)$ at z^{-1} must be equal to 0. So a set of linear algebraic equations is derived

$$\sum_{m=0}^{J-2} \left(G^{(1)}_{jm} C_m + G^{(2)}_{jm} D_m \right) = -g_j, \quad j = 1...2(J-1), \qquad (2.52)$$

in which

$$G^{(1)}_{(2j-1)m} = \int\limits_{c_{2j-1}}^{c_{2j}} \frac{x^m \sin \varphi^*(x)}{f^*(x)\sqrt{|x-b|}} dx, \quad G^{(2)}_{(2j-1)m} = \int\limits_{c_{2j-1}}^{c_{2j}} \frac{x^m \cos \varphi^*(x)}{f^*(x)\sqrt{|x-a|}} dx,$$

$$g_{2j-1} = - \int\limits_{c_{2j-1}}^{c_{2j}} \left[\frac{C_{J-1} + xC_J}{\sqrt{|x-b|}} \sin \varphi^*(x) + \frac{D_{J-1} + xD_J}{\sqrt{|x-a|}} \cos \varphi^*(x) \right] \frac{x^{J-1}}{f^*(x)} dx,$$

$$G^{(1)}_{(2j)m} = \int\limits_{c_{2j-1}}^{c_{2j}} \frac{x^m \cos \varphi^*(x)}{f^*(x)\sqrt{|x-b|}} dx, \quad G^{(2)}_{(2j)m} = - \int\limits_{c_{2j-1}}^{c_{2j}} \frac{x^m \sin \varphi^*(x)}{f^*(x)\sqrt{|x-a|}} dx,$$

$$g_{2j} = -\int_{c_{2j-1}}^{c_{2j}} \left[\frac{C_{J-1} + xC_J}{\sqrt{|x-b|}} \cos\varphi^*(x) - \frac{D_{J-1} + xD_J}{\sqrt{|x-a|}} \sin\varphi^*(x) \right] \frac{x^{J-1}}{f^*(x)} dx,$$

$$j = 1...J-1,$$

where

$$f^*(x) = \sqrt{|x-d|} \prod_{j=1}^{2J-2} \sqrt{|x-c_j|},$$

$$\varphi^*(x) = \text{sgn}(b-a)\varepsilon \ln\left| \frac{\sqrt{\chi(d)} + \sqrt{\chi(x)}}{\sqrt{\chi(d)} - \sqrt{\chi(x)}} \right| +$$

$$+ \varepsilon \sum_{j=1}^{2J-2} (-1)^j \, \text{sgn}(c_j - b) \ln\left| \frac{\sqrt{\chi(c_j)} + \sqrt{\chi(x)}}{\sqrt{\chi(c_j)} - \sqrt{\chi(x)}} \right|, \quad x \in M. \tag{2.53}$$

A solution of Eq. (2.52) can be presented as

$$\begin{cases} C_j = \tilde{p}\left(c_j^{(1)} \cos(\alpha_0 + \beta) + c_j^{(2)} \sin(\alpha_0 + \beta) \right), \\ D_j = \tilde{p}\left(d_j^{(1)} \cos(\alpha_0 + \beta) + d_j^{(2)} \sin(\alpha_0 + \beta) \right), \end{cases} \tag{2.54}$$

where

$$c_{j-1}^{(k)} = \sum_{m=1}^{2J-2} M_{jm}^* h_m^{(k)}, \quad d_{j-1}^{(k)} = \sum_{m=1}^{2J-2} M_{(j+J-1),m}^* h_m^{(k)}, \quad \mathbf{M}^* = \mathbf{M}^{-1},$$

$$M_{j(m+1)} = G_{jm}^{(1)}, \quad M_{j(m+J)} = G_{jm}^{(2)}, \quad m = 0...(J-2), \quad j = 1...(J-1),$$

$$h_{2j}^{(1)} = \int_{c_{2j-1}}^{c_{2j}} \left[\frac{\upsilon - x}{\sqrt{|x-b|}} \cos\varphi^*(x) - \frac{\alpha_1}{\sqrt{|x-a|}} \sin\varphi^*(x) \right] \frac{x^{J-1}}{f^*(x)} dx,$$

$$h_{2j}^{(2)} = \int_{c_{2j-1}}^{c_{2j}} \left[\frac{\alpha_1}{\sqrt{|x-b|}} \cos\varphi^*(x) + \frac{\upsilon + \eta - x}{\sqrt{|x-a|}} \sin\varphi^*(x) \right] \frac{x^{J-1}}{f^*(x)} dx,$$

$$h_{2j-1}^{(1)} = \int_{c_{2j-1}}^{c_{2j}} \left[\frac{\upsilon - x}{\sqrt{|x-b|}} \sin\varphi^*(x) + \frac{\alpha_1}{\sqrt{|x-a|}} \cos\varphi^*(x) \right] \frac{x^{J-1}}{f^*(x)} dx,$$

$$h_{2j-1}^{(2)} = \int_{c_{2j-1}}^{c_{2j}} \left[\frac{\alpha_1}{\sqrt{|x-b|}} \sin\varphi^*(x) - \frac{\upsilon+\eta-x}{\sqrt{|x-a|}} \cos\varphi^*(x) \right] \frac{x^{J-1}}{f^*(x)} dx.$$

In order that the obtained solution will be physically sound, the following additional conditions must be satisfied:

$$\langle u_2'(b) \rangle = 0; \quad \sigma_{22}^{(1)}(x,0) \le 0, \quad x \in L; \quad \langle u_2(x) \rangle \ge 0, \quad x \in M. \tag{2.55}$$

The meaning of these relations is explained in the Sect. 2.1.4. To verify the conditions (2.55), it is necessary to derive the expressions for the stresses and displacements along the interface. Using Eqs. (2.40) and (2.49), they take the form

$$\sigma_{22}^{(1)}(x,0) = (-1)^N \frac{2ge^{\pi\varepsilon}}{f^*(x)} \left(-\mathrm{sgn}(b-a)\frac{P(x)}{\sqrt{|x-b|}} \cosh(\tilde\varphi(x) - \pi\varepsilon) \right.$$
$$\left. + \frac{Q(x)}{\sqrt{|x-a|}} \sinh(\tilde\varphi(x) - \pi\varepsilon) \right), \quad x \in L, \tag{2.56}$$

$$\sigma_{22}^{(1)}(x,0) - i\sigma_{12}^{(1)}(x,0) =$$
$$= -(-1)^N \frac{g(1+\gamma)e^{i\varphi(x)}}{f^*(x)} \left(\frac{P(x)}{\sqrt{|x-b|}} + i\frac{Q(x)}{\sqrt{|x-a|}} \right), \quad x \in U, \tag{2.57}$$

$$\langle u_2'(x) \rangle =$$
$$= -(-1)^N \frac{2\cosh(\pi\varepsilon)}{f^*(x)} \left(\frac{P(x)}{\sqrt{|x-b|}} \cos(\varphi^*(x)) - \frac{Q(x)}{\sqrt{|x-a|}} \sin(\varphi^*(x)) \right), \quad x \in M, \tag{2.58}$$

where N - the number of cracks that are on the right side of x,

$$\tilde\varphi(x) = -2\varepsilon \arctan\sqrt{\frac{\chi(d)}{|\chi(x)|}} -$$
$$- 2\varepsilon\,\mathrm{sgn}(b-a) \sum_{j=1}^{2J-2} (-1)^j \,\mathrm{sgn}(c_j - b) \arctan\sqrt{\frac{\chi(c_j)}{|\chi(x)|}}, \quad x \in L. \tag{2.59}$$

Consider the behavior of normal stresses and the derivative of the displacement jump in the vicinity of the point b. From formulae (2.53) and (2.59) follows that

$$\tilde\varphi(x) = \pi\varepsilon + \varphi_0(b)\sqrt{|x-b|} + O((x-b)^{3/2}), \quad x \to b - 0\,\mathrm{sgn}(b-a), \tag{2.60}$$

$$\varphi^*(x) = \mathrm{sgn}(b-a)\varphi_0(b)\sqrt{|x-b|} + O((x-b)^{3/2}), \quad x \to b + 0\,\mathrm{sgn}(b-a), \tag{2.61}$$

where

$$\varphi_0(b) = \frac{2}{\sqrt{|b-a|}} \left[\frac{1}{\chi(d)} + \mathrm{sgn}(b-a) \sum_{j=1}^{2J-2} (-1)^j \, \mathrm{sgn}(c_j - b) \frac{1}{\sqrt{\chi(c_j)}} \right].$$

Using relations (2.60) and (2.61) and expanding representations (2.56) and (2.58) into Taylor series in the vicinity of the point b, the following formulae are obtained

$$\sigma_{22}^{(1)}(x) = \frac{2(-1)^N g e^{\pi \varepsilon}}{f^*(b)} \left(-\frac{\mathrm{sgn}(b-a)P(b)}{\sqrt{|x-b|}} + \left(-\mathrm{sgn}(b-a)P'(b) + \frac{1}{2}\,\mathrm{sgn}(b-a) \times \right. \right.$$

$$\left. \left. \times P(b)\varphi_0^2(b) + \frac{Q(b)}{\sqrt{|b-a|}}\varphi_0(b) \right) \sqrt{|x-b|} \right) + O\big((x-b)^{3/2}\big), \, x \to b - 0\,\mathrm{sgn}(b-a),$$

$$\langle u_2'(x) \rangle \Big|_{x \to b+0\,\mathrm{sgn}(b-a)} = -\frac{1+\gamma}{2g\gamma} \sigma_{22}^{(1)}(x) \Big|_{x \to b-0\,\mathrm{sgn}(b-a)}.$$

Based on the previous formulae, the conditions (2.55) can be rewritten as

$$P(b) = 0, \quad (-1)^N \left(-\mathrm{sgn}(b-a)P'(b) + \frac{Q(b)}{\sqrt{|b-a|}}\varphi_0(b) \right) < 0.$$

Inserting relations (2.51), (2.54) into (2.50), and then into the above formulae, after some algebraic manipulations follows that

$$\tan(\alpha_0 + \beta) = \frac{(b-\upsilon)b^{J-1} + \sum_{j=0}^{J-2} c_j^{(1)} b^j}{\alpha_1 b^{J-1} - \sum_{j=0}^{J-2} c_j^{(2)} b^j}, \quad \frac{\tan(\alpha_0 + \beta)}{\mathrm{sgn}(p)} > \frac{q_0^{(1)} + q_1^{(1)}}{q_0^{(2)} + q_1^{(2)}},$$

where

$$q_0^{(1)} = \mathrm{sgn}(b-a) \left(b^{J-2}(\upsilon + J(b-\upsilon)) + \sum_{j=1}^{J-2} j c_j^{(1)} b^{j-1} \right),$$

$$q_1^{(1)} = \left(-\alpha_1 b^{J-1} + \sum_{j=1}^{J-2} d_j^{(1)} b^{j-1} \right) \frac{\varphi_0(b)}{\sqrt{|b-a|}},$$

$$q_0^{(2)} = \mathrm{sgn}(b-a) \left(\alpha_1 (J-1)b^{J-2} + \sum_{j=1}^{J-2} j c_j^{(2)} b^{j-1} \right),$$

$$q_1^{(2)} = \left(b^{J-1}(\upsilon + \eta - b) + \sum_{j=1}^{J-2} d_j^{(2)} b^j \right) \frac{\varphi_0(b)}{\sqrt{|b-a|}}.$$

Taking into account (2.51), (2.54) and (2.57), the expressions for the SIFs at the crack tip a, which are determined by the formula

$$K_1 - iK_2 = \lim_{x \to a} (\sigma_{22}(x, 0) - i\sigma_{12}(x, 0))\sqrt{|x - a|},$$

take the form

$$K_1 = 0, \quad K_2 = (-1)^N g(1 + \gamma)Q(a)/f^*(a)$$

or

$$K_2^* \equiv \frac{K_2}{p} = (-1)^N \frac{\cos(\alpha_0 + \beta)}{f^*(a)} \times$$

$$\times \left(-\alpha_1 a^{J-1} + \sum_{j=0}^{J-2} d_j^{(1)} a^j + \left[(\upsilon + \eta - a)a^{J-1} + \sum_{j=0}^{J-2} d_j^{(2)} a^j \right] \tan(\alpha_0 + \beta) \right).$$

2.2.2 Investigation of the Interaction of Two Cracks with Contact Zones

Analyze an interaction of two cracks with lengths l_1 and l_2. The distance between the cracks is denoted by h^*. For the investigation of the cracks interaction it is enough to consider the influence of the relative crack length (l_1/l_2) and the relative distance (h^*/l_1) between them on the relative contact zone length ($\lambda = |b - a|/l_1$) and the SIFs at both tips of a single crack (of length l_1) for various values of the loading parameter (β) and the elastic parameter of materials (ε).

The influence of the distance between the cracks on the contact zone length (λ) of the left crack (l_1) is shown in Figs. 2.10 (for the left contact zone) and 2.11 (for the right one). Here the cracks are assumed to be equal in length ($l_1 = l_2$) and $\beta = \pm\pi/4$.

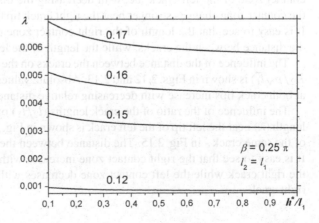

Fig. 2.10 Dependencies of the relative contact zone length near the left tip of the crack on the relative distance between the cracks

Fig. 2.11 Dependencies of the relative contact zone length near the right tip of the crack on the relative distance between the cracks

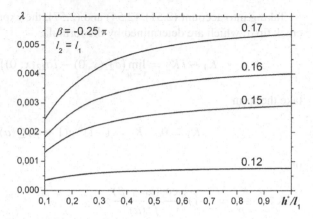

Fig. 2.12 Dependencies of the dimensionless SIF at the left tip of the crack on the relative distance between the cracks

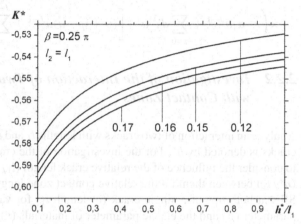

The numbers over the curves indicate the values of the parameter ε for which they are plotted. As it is expected, the right crack (l_2) has more influence on the near (right) contact zone of the left crack, i.e. with decreasing the distance between the cracks the contact zone changes stronger near the right crack tip than near the left crack tip. It is easy to see that the length of the right contact zone decreases with decreasing the distance between the cracks, while the length of the left one increases.

The influence of the distance between the cracks on the dimensionless SIF ($K^* = K_2/p\sqrt{l_1}$) is shown in Figs. 2.12 and 2.13. Absolute values of the dimensionless SIF at both crack tips increase with decreasing relative distance (h^*/l_1).

The influence of the ratio of the crack lengths (l_2/l_1) on the relative contact zone length (λ) near the left tip of the left crack is shown in Fig. 2.14, and near the right tip of the same crack - in Fig. 2.15. The distance between the cracks is fixed ($h^* = l_1$). It is easy to see that the right contact zone increases with decreasing length (l_2) of the right crack while the left contact zone decreases with decreasing length of the right crack.

Fig. 2.13 Dependencies of the dimensionless SIF at the right tip of the crack on the relative distance between the cracks

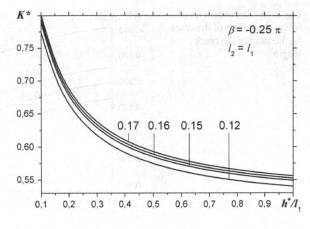

Fig. 2.14 Dependencies of the relative contact zone length near the left tip of the crack on the ratio of the cracks length

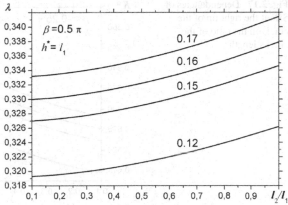

Fig. 2.15 Dependencies of the relative contact zone length near the right tip of the crack on the ratio of the cracks length

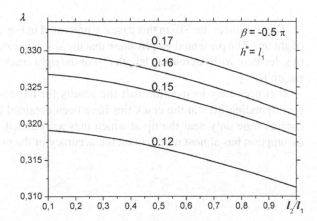

Fig. 2.16 Dependencies of SIF at the left tip of the crack on the ratio of the cracks length

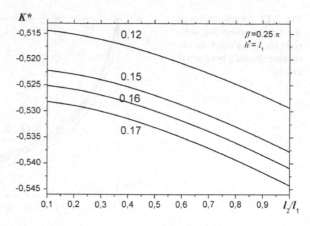

Fig. 2.17 Dependencies of SIF at the right tip of the crack on the ratio of the cracks length

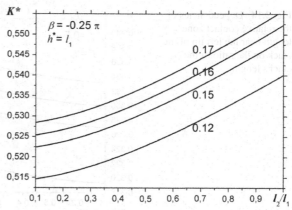

The results for the SIF in this case are presented in Fig. 2.16 (left tip) and Fig. 2.17 (right tip). The presented graphs show that the absolute values of the SIF at both crack tips decrease with decreasing length (l_2) of the right crack in relation to the left crack length (l_1).

It is necessary to mention that the results for the contact zone length and the corresponding SIFs at the crack tips have been obtained by taking into account the contact zone only near the tip at which they are defined. As previously noted, this assumption has almost no effect on the accuracy of the obtained results.

2.3 Analysis of Results and Conclusions

A solution to the plane problem of elasticity for a periodic set of interface cracks with zones of smooth contact of their faces is obtained in a closed form in Sect. 2.1. The dependence of the contact zone length and the SIFs on the distance between the cracks, mechanical parameters of a bimaterial and an external tensile-shear loading is found. The crack growth resistance of the bimaterial decreases

- with decreasing distance between the cracks,
- with increase of the parameter ε which characterize the differences of the elastic properties between the "upper" and "lower" materials

and almost does not depend on the inclination angle of the external loading.

A solution to the plane problem of elasticity for an arbitrary set of the interface cracks with zones of a smooth contact of their faces is obtained in a closed form in Sect. 2.2. An interaction of two cracks when one of the cracks possesses the contact zone is studied in detail. The behavior of the contact zone and the SIFs are determined depending on the relative position of the cracks and their relative sizes. The crack growth resistance of the composite decreases with decreasing distance between the cracks and increasing difference in their sizes. Since the stress intensity factors near the neighboring tips of two cracks are larger than at the outer tips, the cracks will extend towards each other.

References

1. Kozinov S, Loboda V, Kharun I (2006) Periodic system of interface cracks with contact zones in the isotropic bimaterial in the fields of tension and shear. Mater Sci 42(4):533–542
2. Comninou M (1977) The interface crack. J Appl Mech 44, 631–636
3. Gautesen A, Dundurs J (1988) The interface crack under a combined loading. ASME J Appl Mech 55:580–586
4. Kharun I, Loboda V (2003) A set of interface cracks with contact zones in combined tension-shear field. Acta Mechanica 166:43–56
5. Kharun I, Loboda V (2004) A thermoelastic problem for interface cracks with contact zones. Int J Solids Struct 41:159–175
6. Muskhelishvili N (1977) Some basic problems of the mathematical theory of elasticity. Springer, Dordrecht
7. Cherepanov G (1962) About stress state in a heterogeneous plate with slits (in russian). USSR Academy of Sciences. OTN. Mekhanika i mashinostroenie 1, 131–137
8. Gakhov F (1966) Boundary value problems. Pergamon Press, Oxford
9. Nakhmein E, Nuller B (1992) Combined periodic boundary-value problems and their applications in the theory of elasticity. J Appl Math Mech 56:82–89
10. Rice JR, Sih GC (1965) Plane problems of cracks in dissimilar media. J Appl Mech 32:418–423
11. Loboda V (1998) Analytical derivation and investigation of the interface crack models. Int J Solids Struct 35:4477–4489

Chapter 3
Set of Cracks with Contact Zones Located at the Interface of Two Anisotropic Materials

A solution to the problem for a *periodic* set of cracks with contact zones located at the interface of two dissimilar anisotropic materials is constructed in a closed form in Sect. 3.1. By presenting mechanical fields through the piecewise analytical vector functions, a problem is reduced to a homogeneous combined periodic Dirichlet-Riemann boundary value problem, a solution of which is obtained in a closed form. As a result of the numerical analysis of the obtained solution for various combinations of materials, the dependence of the relative contact zone length and the SIF on

- a ratio of the crack length to the period
- an external loading
- a degree of anisotropy

is investigated.

Additionally, a problem for a plane consisting of two dissimilar anisotropic materials with an *arbitrary* set of interface cracks is formulated and its solution is derived in Sect. 3.2. The cracks possess smooth contact zones and are in the field of the uniformly distributed tensile and shear loading applied at infinity. The equations for the determination of the contact zone lengths and the expressions for the SIFs at the crack tips are obtained.

3.1 Periodic Set of Interface Cracks

3.1.1 Statement and Solution of the Problem for a Periodic Set of Cracks with Contact Zones

Statement of the problem. Consider a periodic set of cracks with period h which are located along the interface of two dissimilar half-planes (half-spaces) [1, 2]. Compliance coefficients of the "upper" and "lower" half-planes are denoted by $s_{ij}^{(1)}$ and $s_{ij}^{(2)}$, respectively.

S. Kozinov and V. Loboda, *Fracture Mechanics of Electrically Passive and Active Composites with Periodic Cracking along the Interface*, Springer Tracts in Mechanical Engineering, https://doi.org/10.1007/978-3-030-43138-9_3

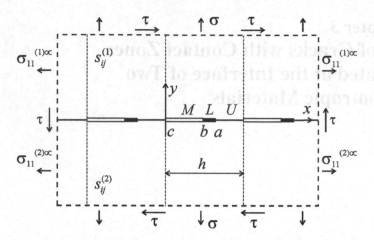

Fig. 3.1 Periodic set of the interface cracks with contact zones between two anisotropic materials

The cracks are opened under the action of the remote tensile (σ) and shear (τ) stresses. As noted earlier, a contact of the crack faces exists in the vicinity of crack tips. Introduce the notations similar to Chap. 2: the point of the crack faces closure is denoted by b, the point between the bonded interface and the contact zone - by a, the point between the bonded interface and the open crack face - by c. In addition, the union of the open crack faces is denoted by M, the contact zones - by L and the bonded parts of the interface - by U (Fig. 3.1). The values related to the "upper" ("lower") material are marked by a superscript "1" ("2").

In this chapter, it is advisable to represent the mechanical fields as a sum of a homogeneous field, that occurs in the plane without cracks under the externally applied mechanical loading, and a field caused by the cracks presence, namely

$$\sigma_{ij}^{(k)}(x, y) = \sigma_{ij}^{(k)\infty}(x, y) + \sigma_{ij}^{*(k)}(x, y).$$

The $\sigma_{ij}^{*(k)}(x, y)$ components are equal to zero at infinity. In this chapter, unless otherwise stated, the mechanical fields are denoted by $\sigma_{ij}^{*(k)}(x, y)$ and a star will be omitted in their denotations. Assume the contact zones to be smooth and the open parts of cracks to be load-free. Let us introduce the stress vector $\mathbf{t} = \{\sigma_{12}, \sigma_{22}\}^{\mathrm{T}}$ and the displacement vector $\mathbf{u} = \{u_1, u_2\}^{\mathrm{T}}$. The continuity and boundary conditions for the mechanical fields disturbed by the cracks can be written in Cartesian coordinates xy, depicted in Fig. 3.1, in the following form which is convenient for further analysis:

$$\langle \mathbf{t}(x) \rangle = 0, \quad x \in L \cup M \cup U, \tag{3.1}$$

$$\langle \mathbf{u}(x) \rangle = 0, \quad x \in U, \tag{3.2}$$

$$\sigma_{12}^{(1)}(x, 0) = -\tau, \quad \langle u_2(x)\rangle = 0, \quad x \in L, \tag{3.3}$$

$$\mathbf{t}^{(1)}(x, 0) = -\mathbf{t}^\infty, \quad x \in M, \tag{3.4}$$

where

$\mathbf{t}^\infty = \{\tau, \sigma\}^T; \sigma_{ij}^{(k)}$ and $u_i^{(k)}$ are stress and displacement components in the "upper" $(k = 1)$ and the "lower" $(k = 2)$ materials,

$$\langle \sigma_{ij}(x)\rangle = \sigma_{ij}^{(1)}(x, +0) - \sigma_{ij}^{(2)}(x, -0), \quad \langle u_i(x)\rangle = u_i^{(1)}(x, +0) - u_i^{(2)}(x, -0)$$

are the jumps of stresses and displacements when passing through the interface of the materials.

It should be noted that the stresses $\sigma_{11}^{(k)\infty}$ should be applied (cf. Fig. 3.1) to satisfy the continuity conditions at infinity [3]. These stresses are determined by the known formulae and have no influence on the parameters of fracture.

Expressions for stresses and a derivative of displacement jump at the interface of materials. Boundary value problem formulation. Under the plane strain or the plane stress conditions, the fields of stresses and displacements in a homogeneous anisotropic medium can be presented in the form [4]:

$$\mathbf{u}(x, y) = \mathbf{A}\mathbf{f}(z) + \overline{\mathbf{A}\mathbf{f}(z)}, \tag{3.5}$$

$$\mathbf{t}(x, y) = \mathbf{B}\mathbf{f}'(z) + \overline{\mathbf{B}\mathbf{f}'(z)}, \tag{3.6}$$

$$\sigma_{11}(x, y) = 2\operatorname{Re}[\mu_1^2 f_1'(z_1) + \mu_2^2 f_2'(z_2)], \tag{3.7}$$

where

$\mathbf{f}(z) = \{f_1(z_1), f_2(z_2)\}^T,$

$f_j(z_j)$ - analytical functions of complex variables $z_j = x + \mu_j y,$

$$\mathbf{A} = \left\| \begin{matrix} s_{11}\mu_1^2 - s_{16}\mu_1 + s_{12} & s_{11}\mu_2^2 - s_{16}\mu_2 + s_{12} \\ s_{12}\mu_1 - s_{26} + s_{22}/\mu_1 & s_{12}\mu_2 - s_{26} + s_{22}/\mu_2 \end{matrix} \right\|,$$

$$\mathbf{B} = \left\| \begin{matrix} -\mu_1 & -\mu_2 \\ 1 & 1 \end{matrix} \right\|,$$

μ_j - complex roots with positive imaginary parts of the equation

$$s_{11}\mu^4 - 2s_{16}\mu^3 + (2s_{12} + s_{66})\mu^2 - 2s_{26}\mu + s_{22} = 0. \tag{3.8}$$

The above expressions are valid for the plane stress. For the plane strain, $s_{ij} - s_{i3}s_{j3}/s_{33}$ should be taken instead of s_{ij}.

For an orthotropic body under the plane stress state, the coefficients s_{ij} are expressed through technical constants in the principal axes as following [4]:

$$s_{11} = \frac{1}{E_1}, \quad s_{12} = -\frac{\nu_{12}}{E_1}, \quad s_{22} = \frac{1}{E_2}, \quad s_{66} = \frac{1}{G_{12}}.$$

In the case of the plane strain, the corresponding coefficients have the form

$$s_{11} = \frac{1 - \nu_{13}\nu_{31}}{E_1}, \quad s_{12} = -\frac{\nu_{12} + \nu_{13}\nu_{32}}{E_1}, \quad s_{22} = \frac{1 - \nu_{23}\nu_{32}}{E_2}, \quad s_{66} = \frac{1}{G_{12}},$$

while all other coefficients are equal to zero.

Substituting Eqs. (3.5) and (3.6) into continuity conditions (3.1) and (3.2), the following formulae can be obtained after elementary transformations

$$\mathbf{B}^{(1)}\mathbf{f}'^{(1)}(x) - \overline{\mathbf{B}}^{(2)}\overline{\mathbf{f}}'^{(2)}(x) = \mathbf{B}^{(2)}\mathbf{f}'^{(2)}(x) - \overline{\mathbf{B}}^{(1)}\overline{\mathbf{f}}'^{(1)}(x), \quad x \in U + M + L,$$

$$\mathbf{A}^{(1)}\mathbf{f}'^{(1)}(x) - \overline{\mathbf{A}}^{(2)}\overline{\mathbf{f}}'^{(2)}(x) = \mathbf{A}^{(2)}\mathbf{f}'^{(2)}(x) - \overline{\mathbf{A}}^{(1)}\overline{\mathbf{f}}'^{(1)}(x), \quad x \in U.$$

Here, any values of the "upper" ("lower") material are denoted by the superscript "1" ("2").

From these equalities follows that

- the vector function $\mathbf{v}(z)$ exist, which is analytical in the entire plane,
- the vector function $\mathbf{w}(z)$ exist, which is analytical in the entire plane except $L \cup M$.

These vector functions are defined by the formulae

$$\mathbf{v}(z) = \begin{cases} \mathbf{B}^{(1)}\mathbf{f}'^{(1)}(z) - \overline{\mathbf{B}}^{(2)}\overline{\mathbf{f}}'^{(2)}(z), & y > 0, \\ \mathbf{B}^{(2)}\mathbf{f}'^{(2)}(z) - \overline{\mathbf{B}}^{(1)}\overline{\mathbf{f}}'^{(1)}(z), & y < 0, \end{cases} \tag{3.9}$$

$$\mathbf{w}(z) = \begin{cases} \mathbf{A}^{(1)}\mathbf{f}^{(1)}(z) - \overline{\mathbf{A}}^{(2)}\overline{\mathbf{f}}^{(2)}(z), & y > 0, \\ \mathbf{A}^{(2)}\mathbf{f}^{(2)}(z) - \overline{\mathbf{A}}^{(1)}\overline{\mathbf{f}}^{(1)}(z), & y < 0, \end{cases} \tag{3.10}$$

Application of the Liouville's theorem to the function $\mathbf{v}(z)$ bring us to the conclusion that $\mathbf{v}(z) = 0$. Thus, the following expressions are obtained from Eqs. (3.9) and (3.10):

$$\mathbf{f}'^{(k)}(z) = \mathbf{D}^{(k)}\mathbf{w}'(z), \quad \overline{\mathbf{f}}'^{(k)}(z) = -\overline{\mathbf{D}}^{(k)}\mathbf{w}'(z), \tag{3.11}$$

where

$$\mathbf{D}^{(k)} = \left[\mathbf{A}^{(k)} - \overline{\mathbf{A}}^{(m)}\left(\overline{\mathbf{B}}^{(m)}\right)^{-1}\mathbf{B}^{(k)}\right]^{-1},$$

$$m = \begin{cases} 1, & k = 2, \\ 2, & k = 1. \end{cases}$$

By inserting (3.11) into (3.5)–(3.7), the following expressions are obtained for the mechanical fields through one vector function $\mathbf{w}(z)$:

$$\sigma_{i2}^{(k)}(x, y) = 2\,\mathrm{Re}[B_{ij}^{(k)} D_{jn}^{(k)} w_n'(z_j^{(k)})], \tag{3.12}$$

$$\sigma_{11}^{(k)}(x, y) = 2\,\mathrm{Re}[(\mu_j^{(k)})^2 D_{jn}^{(k)} w_n'(z_j^{(k)})], \tag{3.13}$$

$$u_i^{(k)}(x, y) = 2\,\mathrm{Re}[A_{ij}^{(k)} D_{jn}^{(k)} w_n(z_j^{(k)})], \tag{3.14}$$

where an Einstein summation over the repeated indices j, n within the limits 1, 2 is used.

Letting $y = 0$ in (3.12)–(3.14) and making simple transformations, the following expressions are derived:

$$\mathbf{t}^{(1)}(x, 0) = \mathbf{G}\mathbf{w}'^{+}(x) - \overline{\mathbf{G}}\mathbf{w}'^{-}(x), \tag{3.15}$$

$$\langle \mathbf{u}'(x) \rangle = \mathbf{w}'^{+}(x) - \mathbf{w}'^{-}(x), \tag{3.16}$$

where $\mathbf{G} = \mathbf{B}^{(1)}\mathbf{D}^{(1)}$.

Expressions (3.15), (3.16) can be rewritten in the following form:

$$\mathbf{N}\mathbf{t}^{(1)}(x, 0) = \mathbf{\Phi}^{+}(x) + \mathbf{\Gamma}\mathbf{\Phi}^{-}(x), \tag{3.17}$$

$$\mathbf{S}\langle \mathbf{u}'(x) \rangle = \mathbf{\Phi}^{+}(x) - \mathbf{\Phi}^{-}(x), \tag{3.18}$$

where

$\mathbf{\Phi}(z) = \mathbf{S}\mathbf{w}'(z)$,

$\mathbf{S} = \mathbf{N}\mathbf{G}$,

$\mathbf{\Gamma} = \mathrm{diag}[\gamma, 1/\gamma]$,

$\mathbf{N}$ is a 2x2 matrix with components $N_{11} = -\dfrac{\overline{G}_{21} + \gamma G_{21}}{\overline{G}_{11} + \gamma G_{11}}$, $N_{21} = \overline{N}_{11}$, $N_{j2} = 1$,

γ is a real constant determined by the formula $\gamma = \dfrac{G^* - \sqrt{G^{*2} - \det \mathbf{G}^2}}{\det \mathbf{G}}$,

$G^* = \mathrm{Re}(G_{21}\overline{G}_{12} - G_{11}\overline{G}_{22})$.

It can be shown that $\Phi_2(z) = \gamma \overline{\Phi}_1(z)$, so instead of (3.17), (3.18) it is enough to write

$$\begin{cases} \sigma_{22}^{(1)}(x, 0) + n_* \sigma_{12}^{(1)}(x, 0) = \Phi^+(x) + \gamma \Phi^-(x), \\ s_*\left(\langle u_1'(x)\rangle - n_*\langle u_2'(x)\rangle\right) = \Phi^+(x) - \Phi^-(x), \end{cases} \qquad (3.19)$$

where $s_* = S_{11}$, $n_* = N_{11}$, $\Phi(z) = \Phi_1(z)$.

For an orthotropic bimaterial with the principal axes coinciding with the coordinate axes, the coefficients in Eq. (3.19) are determined by the following formulae:

$$\left(\frac{1+\gamma}{1-\gamma}\right)^2 = \frac{g_1 g_2}{g_3^2}, \quad n_* = -i\sqrt{\frac{g_2}{g_1}}, \quad s_* = \frac{2\sqrt{g_1 g_2}}{s_0(1+\gamma)},$$

where

$$g_1 = \sum_{k=1}^{2} \mu_*^{(k)} \sqrt{s_{22}^{(k)}},$$

$$g_2 = \sum_{k=1}^{2} \mu_*^{(k)} \sqrt{s_{11}^{(k)}},$$

$$g_3 = \sum_{k=1}^{2} (-1)^k \left(s_{12}^{(k)} + \sqrt{s_{11}^{(k)} s_{22}^{(k)}}\right),$$

$$s_0 = \sum_{k=1}^{2} \sqrt{\frac{s_{22}^{(k)}}{s_{11}^{(k)}}} \prod_{k=1}^{2} \mu_*^{(k)} \sqrt{s_{11}^{(k)}} + \sum_{k=1}^{2} \sqrt{s_{11}^{(k)} s_{22}^{(k)}} \left(2s_{12}^{(m)} + s_{66}^{(k)} + \sqrt{s_{11}^{(m)} s_{22}^{(m)}}\right)$$
$$+ s_{11}^{(2)} s_{22}^{(2)} - (s_{12}^{(1)} - s_{12}^{(2)})^2,$$

$$\mu_*^{(k)} = \sqrt{2s_{12}^{(k)} + s_{66}^{(k)} + 2\sqrt{s_{22}^{(k)} s_{11}^{(k)}}},$$

$$\varepsilon \equiv \frac{\ln \gamma}{2\pi} = \frac{1}{\pi} \operatorname{arctanh}\left(\frac{\sqrt{g_1 g_2}}{g_3}\right).$$

Note that for an isotropic bimaterial the coefficients in Eq. (3.19) are determined by the formulae

$$\gamma = \frac{\mu^{(1)} + \mu^{(2)} \kappa^{(1)}}{\mu^{(2)} + \mu^{(1)} \kappa^{(2)}}, \quad s_* = \frac{2\mu^{(1)} \mu^{(2)}}{\mu^{(1)} + \mu^{(2)} \kappa^{(1)}}, \quad n_* = -i,$$

where

μ - shear modulus,

$$\kappa = \begin{cases} 3 - 4\nu & \text{- for the plane strain,} \\ (3 - \nu)/(1 + \nu) & \text{- for the plane stress,} \end{cases}$$

ν - Poisson's ratio.

Satisfying the boundary conditions (3.3), (3.4) by using Eq. (3.19), the boundary value problem is obtained

$$\begin{cases} F^+(x) + \gamma F^-(x) = 0, & x \in M, \\ \text{Im } F^\pm(x) = 0, & x \in L. \end{cases} \tag{3.20}$$

Here the new function $F(z)$ is introduced:

$$F(z) = \Phi(z) + \tilde{p}e^{i\beta^*}, \tag{3.21}$$

in which

$$\tilde{p} = \frac{p}{1 + \gamma}, \quad p = \sqrt{(\sigma + n'\tau)^2 + (n''\tau)^2}, \quad \beta^* = \arctan\left(\frac{n''\tau}{\sigma + n'\tau}\right),$$

$$n' = \text{Re } n_*, \quad n'' = \text{Im } n_*.$$

Thus, the problem is reduced to the determination of only one function $F(z)$. This function is analytical in the entire plane except for $L \cup M$, where conditions (3.20) should be satisfied.

Solution of the boundary value problem. The problem (3.20) is a periodic combined homogeneous Dirichlet-Riemann boundary value problem. According to Sect. 2.1.2, the general solution to (3.20) can be written in the form

$$F(z) = \frac{e^{i\varphi(z)}}{\sqrt{\Xi(z - c)}} \left(\frac{P(z)}{\sqrt{\Xi(z - b)}} + i\frac{Q(z)}{\sqrt{\Xi(z - a)}} \right), \tag{3.22}$$

where

$$\varphi(z) = 2\varepsilon \ln\left(\frac{\sqrt{\Xi(a - b)\Xi(z - c)}}{\sqrt{\Xi(a - c)\Xi(z - b)} + \sqrt{\Xi(b - c)\Xi(z - a)}} \right),$$

$$P(z) = C_1 \cos[\pi(z - a_*)/h] + C_2 \sin[\pi(z - a_*)/h], \quad a_* = (c + b)/2,$$

$$Q(z) = D_1 \cos[\pi(z - b_*)/h] + D_2 \sin[\pi(z - b_*)/h], \quad b_* = (c + a)/2,$$

C_1, C_2, D_1, D_2 are arbitrary real constants.

The real constants C_1, C_2, D_1, D_2 should be determined from the behavior of the function $F(z)$ at infinity. Analyzing solution (3.22) as $y \to \pm\infty$ similarly to derivations made in Sect. 2.1.2 and taking into account

$$F(z)_{z \to \pm i\infty} = \tilde{p}e^{i\beta^*},$$

the following system of linear algebraic equations is derived

$$\begin{cases} e^{\chi}(\cos\zeta - i\sin\zeta)[C_2 - iC_1 + D_1 + iD_2] = (\sigma + n'\tau + in''\tau)/(1+\gamma), \\ e^{-\chi}(\cos\zeta - i\sin\zeta)[C_2 + iC_1 - D_1 + iD_2] = (\sigma + n'\tau + in''\tau)/(1+\gamma). \end{cases}$$

The real constants χ and ζ are determined in (2.22).
Solving the above system of equations gives

$$\begin{aligned} C_1 &= \sinh\chi[n''\tau\cos\zeta + (\sigma + n'\tau)\sin\zeta]/(1+\gamma), \\ C_2 &= \cosh\chi[(\sigma + n'\tau)\cos\zeta - n''\tau\sin\zeta]/(1+\gamma), \\ D_1 &= -\sinh\chi[(\sigma + n'\tau)\cos\zeta - n''\tau\sin\zeta]/(1+\gamma), \\ D_2 &= \cosh\chi[n''\tau\cos\zeta + (\sigma + n'\tau)\sin\zeta]/(1+\gamma). \end{aligned} \tag{3.23}$$

3.1.2 Derivation of the Classical "Oscillating" Solution as a Particular Case of the "Contact" Model Solution

The solution within the contact model can be analytically reduced to the oscillating one on assumption that the contact zone tends to zero, namely:

$$F(z)\Big|_{b\to a} = $$
$$= e^{i\cdot 2\varepsilon \ln \frac{\sqrt{\Xi(a-b)\Xi(z-c)}}{\sqrt{\Xi(a-c)\Xi(z-a)}}} \frac{(C_1 + iD_1)\cos\dfrac{\pi(z - \frac{c+a}{2})}{h} + (C_2 + iD_2)\sin\dfrac{\pi(z - \frac{c+a}{2})}{h}}{\sqrt{\Xi(z-c)\Xi(z-a)}}\Bigg|_{b\to a}.$$

Substituting Eq. (3.23) allows to modify the above expression to

$$F(z)\Big|_{b\to a} = $$
$$= \frac{[-i\cos\zeta + \sin\zeta]\sinh\chi\cos\dfrac{\pi(z - \frac{c+a}{2})}{h} + [\cos\zeta + i\sin\zeta]\cosh\chi\sin\dfrac{\pi(z - \frac{c+a}{2})}{h}}{\sqrt{\Xi(z-c)\Xi(z-a)}} \times$$
$$\times\, e^{i\varepsilon \ln \frac{\Xi(a-b)\Xi(z-c)}{4\Xi(a-c)\Xi(z-a)}} \frac{\sigma + n'\tau + in''\tau}{1+\gamma}\Bigg|_{b\to a}.$$

After elementary transformations follows that

$$
F(z)\Big|_{b\to a} = e^{i\varepsilon \ln \frac{\Xi(a-b)\Xi(z-c)}{4\Xi(a-c)\Xi(z-a)}} e^{i\zeta} \frac{-i\sinh\chi\cos\dfrac{\pi(z-\frac{c+a}{2})}{h} + \cosh\chi\sin\dfrac{\pi(z-\frac{c+a}{2})}{h}}{\sqrt{\Xi(z-c)\Xi(z-a)}} \times
$$

$$
\times \frac{\sigma + n'\tau + in''\tau}{1+\gamma}\Bigg|_{b\to a}.
$$

Afterwards, using formula (2.22) and taking into account the behavior of the functions ζ and χ as $b \to a$, the following expression can be obtained

$$
F(z)\Big|_{b\to a} = \left[\frac{\Xi(z-c)}{\Xi(z-a)}\right]^{i\varepsilon} \times
$$

$$
\times \frac{-\sin[i\varepsilon(a-c)]\cos\dfrac{\pi(z-\frac{c+a}{2})}{h} + \cos[i\varepsilon(a-c)]\sin\dfrac{\pi(z-\frac{c+a}{2})}{h}}{\sqrt{\Xi(z-c)\Xi(z-a)}} \frac{\sigma + n'\tau + in''\tau}{1+\gamma}
$$

and, finally,

$$
F(z)\Big|_{b\to a} = \frac{\Xi(z-\frac{c+a}{2} - i\varepsilon(a-c))}{\sqrt{\Xi(z-c)\Xi(z-a)}} \left[\frac{\Xi(z-c)}{\Xi(z-a)}\right]^{i\varepsilon} \frac{\sigma + n'\tau + in''\tau}{1+\gamma}. \tag{3.24}
$$

3.1.3 Determination of the Contact Zone Length and Basic Fracture Parameters

Using Eqs. (3.19), (3.21) together with the solution (3.22) of the BVP (3.20), the following expressions for the stresses and the derivative of the displacement jump are obtained at the interface of materials:

$$
\sigma_{22}^{(1)}(x,0) = \frac{2g e^{\pi\varepsilon}}{\sqrt{\Xi(x-c)}} \left(\frac{P(x)\cosh(\tilde{\varphi}(x)-\pi\varepsilon)}{\sqrt{\Xi(x-b)}} + \frac{Q(x)\sinh(\tilde{\varphi}(x)-\pi\varepsilon)}{\sqrt{\Xi(a-x)}}\right), \quad x \in L, \tag{3.25}
$$

$$
\sigma_{22}^{(1)}(x,0) + n_*\sigma_{12}^{(1)}(x,0) = \frac{(1+\gamma)e^{i\varphi(x)}}{\sqrt{\Xi(x-c)}} \left(\frac{P(x)}{\sqrt{\Xi(x-b)}} + i\frac{Q(x)}{\sqrt{\Xi(x-a)}}\right), \quad x \in U, \tag{3.26}
$$

$$
\langle u_2'(x)\rangle = \frac{2\cosh(\pi\varepsilon)}{s_*n''\sqrt{\Xi(x-c)}} \left(\frac{P(x)\cos(\varphi^*(x))}{\sqrt{\Xi(b-x)}} - \frac{Q(x)\sin(\varphi^*(x))}{\sqrt{\Xi(a-x)}}\right), \quad x \in M, \tag{3.27}
$$

where

$$\tilde{\varphi}(x) = 2\varepsilon \arctan \sqrt{\frac{\Xi(b-c)\Xi(a-x)}{\Xi(a-c)\Xi(x-b)}}, \quad x \in L,$$

$$\varphi^*(x) = 2\varepsilon \ln \frac{\sqrt{\Xi(a-b)\Xi(x-c)}}{\sqrt{\Xi(a-c)\Xi(b-x)} + \sqrt{\Xi(b-c)\Xi(a-x)}}, \quad x \in M.$$

Equations (3.25)–(3.27) can be used for any position of the point b, but the above obtained solution will be physically sound under the condition that the stresses are compressive in the contact zone, and the upper and lower crack faces do not overlap and close smoothly at the point b. These physical conditions can be written mathematically as following:

$$\langle u_2'(b)\rangle = 0; \quad \sigma_{22}^{(1)}(x,0) \le 0, \quad x \in L; \quad \langle u_2(x)\rangle \ge 0, \quad x \in M. \tag{3.28}$$

Using Eq. (3.28$_1$) and expanding the right part of Eq. (3.27) into Taylor series in the vicinity of the point b, the transcendental equation is obtained to determine the contact zone length

$$P(b) = 0. \tag{3.29}$$

The correct selection of the root is provided by satisfaction of the conditions (3.28$_2$) and (3.28$_3$). Equation (3.29) can be rewritten as following:

$$\tan \frac{\pi(b-c)}{2h} = -\frac{n''\tau \cos\zeta + (\sigma + n'\tau)\sin\zeta}{(\sigma + n'\tau)\cos\zeta - n''\tau \sin\zeta} \tanh\chi. \tag{3.30}$$

The SIFs at the crack tip a can be defined by a formula

$$K_1 - iK_2 = \lim_{x\to a}\left(\sigma_{22}(x,0) - i\sigma_{12}(x,0)\right)\sqrt{\Xi(x-a)}.$$

Using Eq. (3.26) the following formulae are obtained:

$$K_1 = -n'K_2,$$

$$K_2 = \frac{1+\gamma}{n''\sqrt{\Xi(a-c)}}\left(D_1 \cos\frac{\pi(a-c)}{2h} + D_2 \sin\frac{\pi(a-c)}{2h}\right) =$$

$$= \frac{1}{n''\sqrt{\Xi(a-c)}}\left\{\begin{array}{l}-[(\sigma + n'\tau)\cos\zeta - n''\tau \sin\zeta]\sinh\chi \cos\dfrac{\pi(a-c)}{2h} + \\[2mm] +[n''\tau \cos\zeta + (\sigma + n'\tau)\sin\zeta]\cosh\chi \sin\dfrac{\pi(a-c)}{2h}\end{array}\right\}. \tag{3.31}$$

It is necessary to note that n' is equal to zero in case when the principal axes coincide with the coordinate axes. Consequently, $K_1 = 0$ in this case.

3.1.4 Assessment of the Obtained Solution and Analysis of the Results

In order to investigate the periodic set of interface cracks, it is necessary to establish the dependence of the relative contact zone length $\lambda = (a - b)/l$ and the SIF at the crack tip a on the crack length l and the inclination angle β of the vector of the resultant loading $\sqrt{\sigma^2 + \tau^2}$ to the y-axis ($\tan \beta = \tau/\sigma$). Positive (negative) values of ε show that the "lower" ("upper") material is more rigid.

For numerical analysis, consider an orthotropic bimaterial with the principal axes of the orthotropy coinciding with the coordinate axes. The corresponding graphs for the bimaterial consisting of fiberglass, formed by a system of two fibers, C-III-15-48 (lower material) and silicon (upper) ($n_* = -0.955i$; $\varepsilon = 0.0867$) are denoted by I, fiberglass Glass/DX210 and carbon fiber reinforced plastic AS/4397 ($n_* = -0.634i$; $\varepsilon = 0.0273$) by II. The mechanical properties of the composite materials [5, 6] are shown in Table 3.1.

The dependence of the relative contact zone length λ on the distance between the cracks under the remote tensile loading is shown in Fig. 3.2. Corresponding results for the dimensionless SIF $K^* = -K_2/(p\sqrt{l})$ are presented in Fig. 3.3. As expected, decrease of the distance between the cracks influences the contact zone length only for l/h close to 1 and leads to the significant growth of the SIF.

The results for the contact zone length under the combined loading ($\sigma/\tau = -1/5$) are presented in Fig. 3.4. A comparison with Fig. 3.2 reveals that the presence of the shear loading leads to an increase of the contact zone. Application of the shear loading in the direction shown in Fig. 3.1 leads to the formation of the essentially larger contact zone in the vicinity of the investigated crack tip than in the vicinity of the left tip.

A comparison of the normal displacement jump is shown in Fig. 3.5. The oscillating and contact models are considered for an interface crack when approaching to the right crack tip under the combined ($\sigma = 1$ MPa, $\tau = -2$ MPa) loading. Only small

Table 3.1 Mechanical properties of the composite material components

	Fiberglass C-III-15-48	Silicone	Fiberglass glass/DX210	Carbon fiber AS/4397
E_x, GPa	21.3	169.1	37.8	128.5
E_y, GPa	17.9	130.1	10.1	9.78
G_{xy}, GPa	3.87	50.9	4.9	5.42
v_{xy}	0.16	0.36	0.29	0.3

neighborhood of the right crack tip is shown. Here, the behavior of displacements
for both models is different. However, the displacement fields begin to coincide in a
further distance from the tip.

Figure 3.6 shows a comparison of the normal stress fields (in MPa) on the extension
of the crack along the interface of materials obtained within the oscillating and
contact models under the combined loading. The composite consists of a fiberglass,
formed by a system of two fibers, C-III-15-48 (lower material) and a silicon (upper)
($n_* = -0.955i$; $\varepsilon = 0.0867$). The ratio of the tensile loading to the shear one is
equal to -2, the crack length is half of a period, the relative contact zone length is
equal to 2.07204×10^{-3}, $x_1' = (x_1 - a)/l$. From the results follows that the stresses
are essentially the same starting from a distance of about seven contact zone lengths.
The comparison of the tangential stress fields is shown in Fig. 3.7.

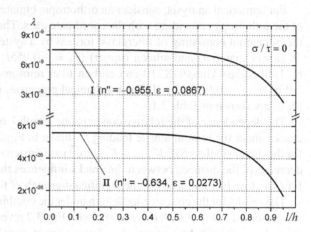

Fig. 3.2 Dependence of the
relative contact zone length λ
on the distance between the
cracks under tensile loading

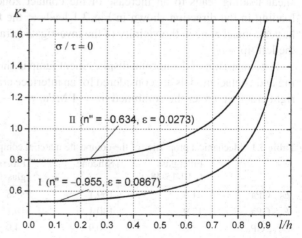

Fig. 3.3 Dependence of the
dimensionless SIF
$K^* = -K_2/(p\sqrt{l})$ on the
distance between the cracks
under tensile loading

Fig. 3.4 Dependence of the relative contact zone length λ on the distance between the cracks under combined loading

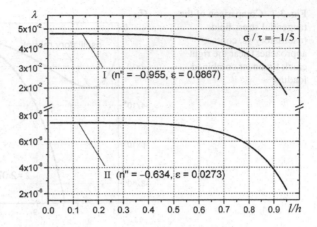

Fig. 3.5 Comparison of the normal displacement jump (oscillatory vs. contact approaches) near the right crack tip under combined loading

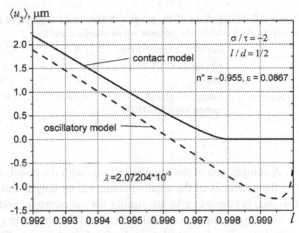

Fig. 3.6 Comparison of the normal stress fields (oscillatory vs. contact approaches) near the right crack tip under combined loading

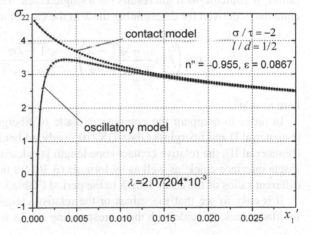

Fig. 3.7 Comparison of the tangential stress fields (oscillatory vs. contact approaches) near the right crack tip under combined loading

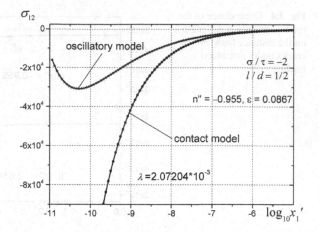

Table 3.2 Comparison of the relative contact zone length and the normalized SIF for a single interface crack and for the periodic set of interface cracks

	Composite components	[7]	$l/h = 1/20$	$l/h = 3/4$
λ	Fiberglass C-III-15-48 - silicone	2.12089×10^{-3}	2.12087×10^{-3}	1.76566×10^{-3}
$-K_2/(p\sqrt{l})$	Fiberglass glass/DX210 - carbon fibre AS/4397	0.7908	0.7909	1.1320

A comparison of the obtained results with the results of the Chap. 2 indicates that the dependences of the relative contact zone length and the SIF on a change of the distance between cracks are similar for the isotropic and anisotropic bimaterials.

It is important to note that the values of the relative contact zone length at small ratio l/h coincide with the results for a single crack. The following transcendental equation is derived for determining the λ in the case of a single crack [7]:

$$\tan\left(\varepsilon \ln \frac{1 - \sqrt{1 - \lambda}}{1 + \sqrt{1 - \lambda}}\right) = \frac{\sqrt{1 - \lambda}\sigma + 2\varepsilon m\tau}{2\varepsilon\sigma - \sqrt{1 - \lambda}m\tau}, \tag{3.32}$$

here $m = n''$.

In order to compare the composites made of fiberglass C-III-15-48 - silicone (bimaterial I) and fiberglass Glass/DX210 - carbon fiber reinforced plastic AS/4397 (bimaterial II), the relative contact zone length is calculated by formula (3.32) for a single interface crack, as well as by formula (3.30) for the periodic set of cracks for different ratios of the crack length to the period (Table 3.2).

It is easy to see that the values of the relative contact zone length for a single interface crack coincide with the corresponding results for the periodic set of cracks at a small ratio of the crack length to the period. This agreement shows that the results obtained for the periodic set of interface cracks are consistent with those for a single interface crack.

3.2 Interaction of an Arbitrary Set of Interface Cracks

3.2.1 Statement and Solution of the Dirichlet-Riemann BVP for a Set of Interface Cracks with Contact Zones

Statement of the problem. Consider a set of cracks located between dissimilar anisotropic half-planes (half-spaces) taking into account the contact of faces in the vicinity of the crack tips. Compliance coefficients of the "upper" and "lower" half-planes are $s_{ij}^{(1)}$ and $s_{ij}^{(2)}$, respectively. The plane is subjected to a uniformly distributed tensile (σ) and shear (τ) loading at infinity.

Introduce the following designations: the point of the crack faces closure is denoted by b, the point between the bonded interface and the contact zone - by a, the point between the bonded interface and the open crack face - by d. Other crack tips are denoted by c_j (cf. Fig. 3.8). Additionally, the union of the open crack faces is denoted by M, the contact zones - by L and the bonded parts of the interface - by U. The contact zones are assumed to be smooth. The open regions of the cracks are load-free. Introducing the stress vector $\mathbf{t} = \{\sigma_{12}, \sigma_{22}\}$ and the displacement vector $\mathbf{u} = \{u_1, u_2\}$, the continuity and boundary conditions for mechanical fields disturbed by the cracks in Cartesian coordinates xy (cf. Fig. 3.8) can be written as formulae (3.1)–(3.4).

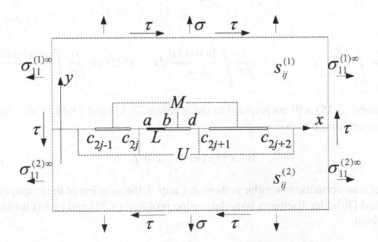

Fig. 3.8 Finite set of the interface cracks between two anisotropic materials

Identical to Sect. 3.1, the fields of stresses and displacements in a homogeneous anisotropic medium under the plane strain or plane stress condition can be expressed in the form (3.5)–(3.7). Conducting transformations similar to (3.9)–(3.18), one can obtain the expressions (3.19) for the stress fields and the derivative of the displacement jump at the material interface.

Satisfying the boundary conditions (3.3), (3.4) by using Eq. (3.19) and introducing the new function according to Eq. (3.21), the following boundary value problem is obtained:

$$F^+(x) + \gamma F^-(x) = 0, \quad x \in M, \tag{3.33}$$

$$\text{Im } F^\pm(x) = 0, \quad x \in L. \tag{3.34}$$

Thus, the problem is reduced to the determination of the only one function $F(z)$ which is analytical in the entire plane except $L \cup M$, where Eqs. (3.33) and (3.34) should be satisfied.

General analytical solution. The boundary value problem similar to (3.33) and (3.34) was considered in [8], where a solution to the problem of pressing a rigid punch into an isotropic half-plane taking into account the detachment zones was obtained. However, in our case the problem can be solved by a neater method. Let us replace the BVP (3.33) and (3.34) by the following linear conjugation problem:

$$X^+(x) - \begin{Bmatrix} G(x) \\ -\gamma \end{Bmatrix} X^-(x) = 0, \quad x \in \begin{Bmatrix} L \\ M \end{Bmatrix},$$

a solution of which can be written in the form according to [9]:

$$X(z) = e^{\psi(z)}, \quad \psi(z) = \left(\frac{1}{2} - i\varepsilon \right) I(z) - i\Gamma_1(z) + \Gamma_2(z), \tag{3.35}$$

where

$$I(z) = \int_M \frac{dx}{x - z}, \quad \Gamma_1(z) = \frac{1}{2\pi} \int_L \frac{\ln|G(x)|}{x - z} dx, \quad \Gamma_2(z) = \frac{1}{2\pi} \int_L \frac{\arg[G(x)]}{x - z} dx.$$

Solution (3.35) will correspond to the problem (3.33) and (3.34) if the function $G(x)$ is such that it satisfies

$$\text{Im } X^\pm(x) = 0, \quad x \in L. \tag{3.36}$$

After transformations similar to those in Chap. 2, the solution of the homogeneous combined Dirichlet-Riemann boundary value problem (3.33) and (3.34) is obtained in the form

$$F(z) = \frac{e^{i\varphi(z)}}{f(z)} \left(\frac{P(z)}{\sqrt{z - b}} + i \frac{Q(z)}{\sqrt{z - a}} \right). \tag{3.37}$$

Here

$$f(z) = \sqrt{(z-d)} \prod_{j=1}^{2J-2} (z - c_j)^{1/2},$$

$P(z)$ and $Q(z)$ are polynomials with arbitrary real coefficients.

From the single-valuedness conditions of displacements [10] and from the condition that the function $F(z)$ must be a constant $\tilde{p}e^{i\beta}$ at infinity according to Eq. (3.21) follows that the polynomials $P(z)$ and $Q(z)$ must have the form

$$P(z) = \sum_{j=0}^{J} C_j z^j, \quad Q(z) = \sum_{j=0}^{J} D_j z^j, \tag{3.38}$$

where real coefficients of the polynomials are determined as

$$\begin{cases} C_j = \tilde{p} \left(c_j^{(1)} \cos(\alpha_0 + \beta) + c_j^{(2)} \sin(\alpha_0 + \beta) \right), \\ D_j = \tilde{p} \left(d_j^{(1)} \cos(\alpha_0 + \beta) + d_j^{(2)} \sin(\alpha_0 + \beta) \right), \end{cases} \quad (j = 0 \ldots J), \tag{3.39}$$

where

$$c_J^{(1)} = -d_J^{(2)} = 1, \quad c_J^{(2)} = d_J^{(1)} = 0, \quad c_{J-1}^{(2)} = d_{J-1}^{(1)} = -\alpha_1, \quad c_{J-1}^{(1)} = -v, \quad d_{J-1}^{(2)} = \eta,$$

$$c_{j-1}^{(k)} = \sum_{m=1}^{2J-2} M_{jm}^* h_m^{(k)}, \quad d_{j-1}^{(k)} = \sum_{m=1}^{2J-2} M_{(j+J-1)m}^* h_m^{(k)},$$

$$\mathbf{M}^* = \mathbf{M}^{-1}, \quad M_{n(m+1)} = G_{nm}^{(1)}, \quad M_{n(m+J)} = G_{nm}^{(2)},$$

$$v = \frac{1}{2} \left(b + d + \sum_{j=1}^{2J-2} c_j \right), \quad \eta = \frac{1}{2} \left(a + d + \sum_{j=1}^{2J-2} c_j \right),$$

and values $G_{(2j-1)m}^{(1)}$, $G_{(2j)m}^{(1)}$, $G_{(2j-1)m}^{(2)}$, $G_{(2j)m}^{(2)}$, $h_{2j}^{(1)}$, $h_{2j}^{(2)}$, $h_{2j-1}^{(1)}$, $h_{2j-1}^{(2)}$, α_0, α_1 are determined at Sect. 2.2.1.

Functions $\varphi^*(x)$, $f^*(x)$, $x \in M$ were determined earlier in Chap. 2 by expressions (2.53),

$$m = 0 \ldots (J-2), j = 1 \ldots (J-1), n = 1 \ldots (2J-2),$$

J - total number of cracks.

Determination of the contact zones length and the SIF. For the obtained solution to be physically sound, it is necessary that the following mathematical conditions are satisfied:

$$\langle u_2'(b) \rangle = 0; \quad \sigma_{22}^{(1)}(x, 0) \le 0, \quad x \in L; \quad \langle u_2(x) \rangle \ge 0, \quad x \in M. \tag{3.40}$$

The explanation of the meaning of these conditions was given in the previous chapter.

To verify the conditions (3.40), it is necessary to derive the expressions for the stresses and displacements along the x-axis. Consider the behavior of stresses and displacement jump in the vicinity of the point b. Inserting the solution of the BVP (3.37) into the expression for the new function $F(z)$ (3.21) and then into the expression for the stresses and displacement jump along the interface (3.19) via the function $\Phi(z)$ gives

$$\sigma_{22}^{(1)}(x, 0) =$$
$$= \frac{2e^{\pi\varepsilon}}{(-1)^N f^*(x)} \left(\frac{\delta_0 P(x)}{\sqrt{|x-b|}} \cosh(\tilde{\varphi}(x) - \pi\varepsilon) + \frac{Q(x)}{\sqrt{|x-a|}} \sinh(\tilde{\varphi}(x) - \pi\varepsilon) \right), \quad x \in L,$$
(3.41)

$$\sigma_{22}^{(1)}(x, 0) + n_*\sigma_{12}^{(1)}(x, 0) = -(-1)^N e^{i\varphi(x)} \frac{(1+\gamma)}{f^*(x)} \left(\frac{P(x)}{\sqrt{|x-b|}} + \frac{Q(x)}{\sqrt{|x-a|}} \right), \quad x \in U,$$
(3.42)

$$\langle u_2'(x) \rangle = \frac{2(-1)^N \cosh(\pi\varepsilon)}{s_* n'' f^*(x)} \left(\frac{P(x)}{\sqrt{|x-b|}} \cos(\varphi^*(x)) - \frac{Q(x)}{\sqrt{|x-a|}} \sin(\varphi^*(x)) \right). \, x \in M,$$
(3.43)

Here N - the number of cracks located to the right of x,
$$\tilde{\varphi}(x) = -2\varepsilon \arctan\sqrt{\chi(d)/|\chi(x)|} + 2\varepsilon\delta_0 \sum_{j=1}^{2J-2} (-1)^j \delta_j \arctan\sqrt{\chi(c_j)/|\chi(x)|},$$
$$x \in L,$$
$$\delta_0 = \mathrm{sgn}(a - b), \quad \delta_j = \mathrm{sgn}(c_j - b).$$

Expanding Eqs. (3.41) and (3.43) into Taylor series in the vicinity of the point b, the following expressions for the mechanical field are obtained:

$$\sigma_{22}^{(1)}(x) = \frac{2e^{\pi\varepsilon}}{(-1)^N f^*(b)} \left(\frac{\delta_0 P(b)}{\sqrt{|x-b|}} + \varphi_1\sqrt{|x-b|} \right) + O\big((x-b)^{3/2}\big), \quad x \to b + 0\delta_0,$$
(3.44)

$$\langle u_2'(x) \rangle \big|_{x \to b - 0\delta_0} = -\frac{1+\gamma}{2s_* n'' \gamma} \sigma_{22}^{(1)}(x) \bigg|_{x \to b + 0\delta_0},$$
(3.45)

where

$$\varphi_1 = \delta_0 P'(b) - \frac{\delta_0}{2} P(b)\varphi_0^2 + \frac{Q(b)}{\sqrt{|b-a|}}\varphi_0,$$

$$\varphi_0 = \frac{2}{\sqrt{|b-a|}} \left[\frac{1}{\chi(d)} - \delta_0 \sum_{j=1}^{2J-2} (-1)^j \delta_j \frac{1}{\sqrt{\chi(c_j)}} \right].$$

Finally, the condition (3.40) can be written as

$$P(b) = 0, \quad (-1)^N \varphi_1 < 0.$$

If the stresses are compressive in the contact zone then from Eq. (3.45) follows that the crack faces do not interpenetrate each other in the vicinity of the closure point.

Inserting (3.39) into (3.38) and then into the above formulae, the following expressions can be written after some algebraic manipulations:

$$\tan(\alpha_0 + \beta) = \frac{(b - v)b^{J-1} + \sum_{j=0}^{J-2} c_j^{(1)} b^j}{\alpha_1 b^{J-1} - \sum_{j=0}^{J-2} c_j^{(2)} b^j}, \quad \frac{\tan(\alpha_0 + \beta)}{\text{sgn}(p)} > \frac{q_0^{(1)} + q_1^{(1)}}{q_0^{(2)} + q_1^{(2)}}, \quad (3.46)$$

where

$$q_0^{(1)} = -\delta_0 \left(b^{J-2}(v + J(b - v)) + \sum_{j=1}^{J-2} j c_j^{(1)} b^{j-1} \right),$$

$$q_1^{(1)} = \left(-\alpha_1 b^{J-1} + \sum_{j=1}^{J-2} d_j^{(1)} b^{j-1} \right) \frac{\varphi_0}{\sqrt{|b - a|}},$$

$$q_0^{(2)} = -\delta_0 \left(\alpha_1 (J - 1) b^{J-2} + \sum_{j=1}^{J-2} j c_j^{(2)} b^{j-1} \right),$$

$$q_1^{(2)} = \left(b^{J-1}(\eta - b) + \sum_{j=1}^{J-2} d_j^{(2)} b^j \right) \frac{\varphi_0}{\sqrt{|b - a|}}.$$

The transcendental equation (3.46_1) determines the position of the point b or the value of the contact zone at which the crack faces close smoothly. However, the transcendental equation has several solutions and it is necessary to choose those, for which contact stresses are compressive in the vicinity of the point b and the crack faces do not penetrate each other. It can be shown that the inequality (3.46_2) uniquely identifies the required root of the Eq. (3.46_1).

Inserting (3.39) into (3.38) and subsequently into (3.41), the expressions for the SIFs at the crack tip a, which are determined by the formula

$$K_1 - iK_2 = \lim_{x \to a} \left(\sigma_{22}(x, 0) - i\sigma_{12}(x, 0) \right) \sqrt{|x - a|},$$

after algebraic manipulations take the form

$$K_1 = -n'K_2, \quad \frac{K_2}{p} = -(-1)^N \frac{\cos(\alpha_0 + \beta)}{n'' f^*(a)} (\omega_1 + \omega_2 \tan(\alpha_0 + \beta)), \quad (3.47)$$

where

$$\omega_1 = -\alpha_1 a^{J-1} + \sum_{j=0}^{J-2} d_j^{(1)} a^j, \quad \omega_2 = (\eta - a) a^{J-1} + \sum_{j=0}^{J-2} d_j^{(2)} a^j.$$

In the case of a single crack of the length l, Eq. (3.46) for determining the relative contact zone length λ takes the form

$$\arctan\left(\frac{\sqrt{1-\lambda}}{2\varepsilon} + 2\varepsilon \, \text{arctanh}\left(\sqrt{1-\lambda}\right)\right) = \pm\beta + \pi. \tag{3.48}$$

Using Eq. (3.48), the expression (3.47$_2$) for the SIF takes the following form:

$$\frac{K_2}{p\sqrt{l}} = \pm\frac{1+4\varepsilon^2}{2n''}\left(1 + \frac{4\varepsilon^2}{1-\lambda}\right)^{-1/2}. \tag{3.49}$$

The "+" sign in Eqs. (3.48), (3.49) is taken for the determination of the contact zone and the SIF in the vicinity of the right crack tip, and the "−" sign is related to the left crack tip. Due to the fact that the contact zone lengths are small, Eqs. (3.48) and (3.49) can be expanded into a power series for λ that leads to the following elementary formulae:

$$\lambda = 2\left(\frac{1}{1-t} + \frac{1+t}{1+4\varepsilon^2}\right)^{-1} + O(\lambda^2), \quad t = \tanh\left[\frac{1}{2\varepsilon}\left(\pm\beta - \arctan\left(\frac{1}{2\varepsilon}\right) + \pi\right)\right], \tag{3.50}$$

$$\frac{K_2}{p\sqrt{l}} = \frac{\pm1}{n''}\left(\frac{1}{2}\sqrt{1+4\varepsilon^2} - \frac{\varepsilon^2}{\sqrt{1+4\varepsilon^2}}\lambda\right) + O(\lambda^2). \tag{3.51}$$

3.2.2　Investigation of the Interaction of Two Cracks with Contact Zones

In order to define fundamental patterns of the behavior of the SIF and the size of the contact zones for a set of interface cracks, it is enough to investigate the interaction of two cracks. Denote the crack lengths as l_1 and l_2 and the distance between them as h^*. Consider the behavior of the SIF and the relative contact zone lengths of the crack l_1 depending on the dimensionless parameters l_2/l_1 and h^*/l_1. As for the other parameters that appear in the solution (s_*, n_*, β and ε):

- the parameter s_* is not included in the expression for the SIF and the relative contact zone length,
- the parameter $n_* = n' + in''$ is included in the expression for the SIF as a coefficient, so its influence is obvious,

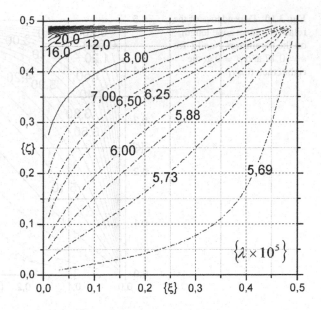

Fig. 3.9 Dependence of the relative contact zone length on the parameters ξ and ζ for $\varepsilon = 0.17$ and $\beta = 0$ at the left crack tips l_1

- the influence of the parameters ε and β on the SIF and the value of the contact zones is evident from formulae (3.50) and (3.51).

It should be noted that in case of an isotropic bimaterial, the parameter β is an inclination angle of the resulting loading vector p to the y-axis. The following formulae can be written for an anisotropic bimaterial

$$\tan \beta = \frac{n'' \tan \beta_0}{1 + n' \tan \beta_0}, \quad \tan \beta_0 = \frac{\tau}{\sigma}.$$

It follows from the last expression that the values of the relative contact zone length and the SIF (with a multiplier $-n''$) calculated for an anisotropic bimaterial with an inclination angle β_0 correspond to the values calculated for an isotropic bimaterial with an inclination angle β, under the condition that the parameters ε are the same.

The dependences of the relative contact zone lengths are shown in Figs. 3.9 and 3.10 at the left and right crack tips l_1, respectively. The dependences are presented as functions of the parameters $\xi = \frac{1}{\pi} \tan \frac{h^*}{l_1}$ and $\zeta = \frac{1}{\pi} \tan \frac{l_2}{l_1}$ for constant values $\varepsilon = 0.17$ and $\beta = 0$.

Similar dependences for the normalized SIF $n'' K_2 / p \sqrt{l_1}$ are shown in Figs. 3.11 and 3.12 at the left and right crack tips, respectively. The dependences presented in figures remain qualitatively the same while changing the values of the parameters ε and β.

As it is seen from the figures, the relative contact zone length in the vicinity of the left crack tip l_1 increases with decreasing distance between the cracks and

Fig. 3.10 Dependence of the relative contact zone length on the parameters ξ and ζ for $\varepsilon = 0.17$ and $\beta = 0$ at the right crack tips l_1

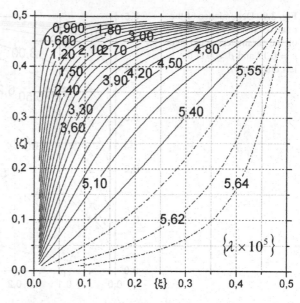

Fig. 3.11 Dependence of the normalized SIF on the parameters ξ and ζ for $\varepsilon = 0.17$ and $\beta = 0$ at the left crack tips l_1

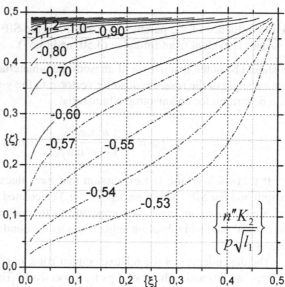

Fig. 3.12 Dependence of the normalized SIF on the parameters ξ and ζ for $\varepsilon = 0.17$ and $\beta = 0$ at the right crack tips l_1

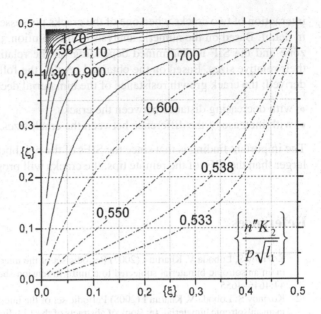

with increasing length of the second crack. On the contrary, the right contact zone decreases under the same conditions. The absolute values of the normalized SIF increase with decreasing h and increasing l_2 at both left and right tips. Besides, the normalized SIF takes larger values at the right crack tip which is closer to the second crack than the left crack tip.

3.3 Analysis of Results and Conclusions

A solution to the plane problem of elasticity for a periodic set of cracks located at the interface of two dissimilar anisotropic materials with zones of smooth contact of the crack faces is obtained in closed form in Sect. 3.1. The dependence of the contact zone length and the SIF on the distance between cracks, mechanical properties of a bimaterial and an external tensile-shear loading is found. Based on the obtained results, the following conclusions are derived: the crack growth resistance of the bimaterial decreases

- with decreasing distance between the cracks,
- with increase of the parameter ε which characterize the differences of the elastic properties between the "upper" and "lower" materials,
- with decrease of the constant n'' which characterizes the degree of the anisotropy

and almost does not depend on the inclination angle of the external loading.

A solution to the plane problem of elasticity for a set of interface cracks with zones of a smooth contact of their faces is obtained in closed form in Sect. 3.2. An interaction of two cracks when one of the cracks possesses a contact zone is studied in detail. Based on the numeric analysis of the solution, the behavior of the contact zone and the SIF is determined depending on the relative position of cracks and their relative sizes. Based on the obtained results, the following conclusions can be derived: the crack growth resistance of the bimaterial decreases

• with decreasing distance between the cracks,
• with increase of the difference between the crack sizes.

The important finding is that since the SIFs of the neighboring tips of two cracks are larger than the SIFs of the remote tips, the cracks will propagate towards each other.

References

1. Kozinov S, Loboda V, Kharun I (2007) Periodic set of the interface cracks with contact zones in an anisotropic bimaterial subjected to a uniform tension-shear loading. Int J Solids Struct 44:4646–4655
2. Kozinov S, Loboda V, Kharun I (2008) Periodic set of the interface cracks with contact zones in an anisotropic bimaterial. In: Book of abstracts of the 17th European conference on Fracture 'Multilevel approach to fracture of materials, componenets and structures'-Brno, p 95
3. Rice JR, Sih GC (1965) Plane problems of cracks in dissimilar media. J Appl Mech 32:418–423
4. Lekhnitsky S (1963) Theory of elasticity of an anisotropic elastic body. San Francisco: Holden-Day
5. Ashkenazi E, Ganov E (1980) Anisotropy of construction materials [in russian]. Mashinostroenie 247
6. Vasiliev V (1990) Composite materials [in russian]. Mashinostroenie
7. Herrmann K, Loboda V (1999) On interface crack models with contact zones situated in an anisotropic bimaterial. Arch Appl Mech 69:317–335
8. Nakhmein E, Nuller B (1988) The pressure of a system of stamps on an elastic half-plane under general conditions of contact adhesion and slip. J Appl Math Mech 52(2):223–230
9. Gakhov F (1966) Boundary value problems. Pergamon Press, Oxford
10. Muskhelishvili N (1977) Some basic problems of the mathematical theory of elasticity. Springer, Dordrecht

Chapter 4
Periodic Set of Cracks Located at the Interface of Piezoelectric Materials

A solution for a periodic set of cracks located at the interface of two dissimilar piezoelectric materials is constructed in a closed form in this chapter.

Some basic information about the piezoelectric materials is highlighted in Sect. 4.1.

Fundamental equations for electromechanical fields in a composite, consisting of two piezoelectric materials, are presented in Sect. 4.2.

Section 4.3 is dedicated to a periodic set of electrically permeable cracks. By presenting electromechanical factors through piecewise analytical vector functions, a problem within the classical model is reduced to a periodic Riemann boundary value problem (Sect. 4.3.1), which is solved in a closed form. Assuming the presence of zones of a smooth contact of faces in the vicinity of the crack tips, the problem is reduced to a combined periodic Dirichlet-Riemann boundary value problem, which is solved in a closed form (Sect. 4.3.2). As a result of the numerical analysis of the obtained solution, the dependence of the relative contact zone length and the stress intensity factors on the ratio of the crack length to the period and on the external electromechanical loading is investigated for various combinations of piezoelectrics. The case of compressive-shear loading, leading to the mainly closed interface, was considered separately in Sect. 4.3.3.

In Sect. 4.4 the problems of cracks with a finite electric permittivity are solved. In Sect. 4.4.1 a solution for a homogeneous piezoelectric material is obtained. In Sect. 4.4.2 a solution for a bimaterial within the classical model, and in Sect. 4.4.3 a solution for a bimaterial taking into account the contact zones in the vicinity of the crack tips is obtained.

4.1 Basic Information About Piezoelectric Materials

Piezoelectric materials are widely used in an engineering practice as sensors, transducers and actuators. However, they are very fragile and inclined to the fracture which can be caused by strong mechanical stresses or electrical fields.

S. Kozinov and V. Loboda, *Fracture Mechanics of Electrically Passive
and Active Composites with Periodic Cracking along the Interface*, Springer Tracts
in Mechanical Engineering, https://doi.org/10.1007/978-3-030-43138-9_4

The piezoelectric effect is an effect of occurrence of dielectric polarization when the mechanical stresses are applied (direct piezoelectric effect). There is the converse piezoelectric effect as well, which is the appearance of the mechanical strains when an electric field is applied. The direct effect was discovered by brothers Jacques and Pierre Curie in 1880. The reverse effect was deduced from fundamental thermodynamic principles by Gabriel Lippmann in 1881, and was experimentally confirmed in the same year by the Curies.

The presence of piezoelectric properties is closely related to the symmetry of the crystals. There are 32 crystal classes but only 20 of them can have piezoelectric effect. Piezoelectric crystals are electrically neutral: the atoms inside the crystal can be arranged unsymmetrically, but their electrical charges are balanced (positive and negative charges cancel themselves). If a piezoelectric crystal is stretched or squeezed, some of the atoms are pushed closer together or further apart, braking the electrical balance and causing net electrical charges to appear. Some of the natural crystals with piezoelectric effect are: Rochelle salt (potassium sodium tartrate), quartz, topaz, tourmaline and other. The naturally occurring piezoelectric crystals are nonrenewable natural resources with natural piezoelectric effect. However, two crystals will have scatter in piezoelectric properties which is unacceptable for the industry requiring very stable electromechanical properties in every production batch. Dry bone exhibits some piezoelectric properties due to collagen. Some other biological materials possessing piezoelectric properties are enamel and dentin, tendon, etc.

Artificially created in the early 50s piezoelectric materials exhibit remarkable properties and in most cases are preferable to natural piezoelectric crystals. An important advantage of them is the ability to be manufactured in different geometric shapes according to the constructive purpose and the possibility to tune the electromechanical properties. Such artificially created piezoelectric materials are called piezoceramics. E.g., a polycrystalline PZT piezoceramic comprising numerous crystallites (Weiss domains) is a solid solution of the lead zirconate titanate. Each domain consists of a plurality of the unit cells of uniform dipole direction. The direction of polarization between neighbouring Weiss domains within a tetragonal crystallite can differ by 90° or 180°. In the initial state, a polycrystalline ceramic has a higher symmetry relative to the symmetry of the individual crystals and is an isotropic dielectric material. The polycrystalline ceramic does not possess a piezoelectric effect at this state. If to subject a separate domain to the mechanical loading, free charges emanate on the boundary surface. The amount of charges in polycrystalline body with many domains equals to zero because of their random orientation. The ceramics can be made piezoelectric by a poling treatment which involves exposing it to a strong electric field at a temperature below the Curie point. Under the action of this electric field, domains will nearly align with the field. The material will also lengthen in the direction of the field. When the electric field is removed, the dipoles will remain partially locked giving the ceramic material a remanent polarization and a remanent deformation (i.e. making it anisotropic).

The production and forming processes used in the manufacture of piezoceramic parts are as diverse as their applications and the associated suitability criteria. Typical shapes include disks, rectangular plates, rods, rings, cylinders, tubes, spheres and semi-spheres.

The direct piezoelectric effect is used as the basis for force, pressure, vibration and acceleration sensors as a load sensitive element, e.g. as a sensing element in microphones, in a contact piezoelectric detonators, in piezoelectric igniters for high voltages generation, etc. The converse piezoelectric effect is used as a basis for actuator and displacement devices, e.g. in high-accuracy positioning systems, such as a positioning system of the needle in a scanning tunnel microscope, an atomic force microscopy scanner or as a positioner for moving of the hard disk head. It is used as well in the ultrasound equipment including medical applications, for the ink ejection in large-format printers, in diesel fuel injectors, piezoelectric motors, etc.

4.2 Fundamental Equations for Electromechanical Fields in Piezoelectric Materials

The constitutive equations for a piezoelectric material can be written in the form

$$\Pi_{iJ} = E_{iJKl} V_{K,l},$$ (4.1)

$$\Pi_{iJ,i} = 0,$$ (4.2)

where

$$V_K = \begin{cases} u_k, & K = 1, 2, 3, \\ \varphi, & K = 4, \end{cases}$$ (4.3)

$$\Pi_{iJ} = \begin{cases} \sigma_{iJ}, & i, J = 1, 2, 3, \\ D_i, & i = 1, 2, 3; \quad J = 4, \end{cases}$$ (4.4)

$$E_{iJKl} = \begin{cases} C_{ijkl}, & J, K = 1, 2, 3, \\ e_{lij}, & J = 1, 2, 3; \quad K = 4, \\ e_{ikl}, & K = 1, 2, 3; \quad J = 4, \\ -\varepsilon_{il}, & J = K = 4. \end{cases}$$ (4.5)

Here $u_k, \varphi, \sigma_{ij}, D_i$ are mechanical displacements, electric potential, stresses and electrical displacements, respectively. In Eqs. (4.1)–(4.5) subscripts written in small letters vary from 1 to 3, and in capital ones - from 1 to 4.

All fields are considered independent of the coordinate x_2. Then the solution of Eqs. (4.1)–(4.2) can be presented as

$$V = a f(z),$$

where
$$z = x_1 + \rho x_3,$$

$$\mathbf{V} = \{u_1, u_2, u_3, \varphi\}^T,$$

$f(z)$ is an arbitrary function,

vector $\mathbf{a} = \{a_1, a_2, a_3, a_4\}^T$ is obtained from the following equation:

$$[\mathbf{Q} + \rho(\mathbf{R} + \mathbf{R}^T) + \rho^2 \mathbf{T}]\mathbf{a} = 0. \tag{4.6}$$

The elements of 4×4 matrices $\mathbf{Q}$, $\mathbf{R}$ and $\mathbf{T}$ are defined as

$$Q_{JK} = E_{1JK1}, \quad R_{JK} = E_{1JK3}, \quad T_{JK} = E_{3JK3}.$$

There is a nontrivial solution to (4.6) if ρ is the root of the equation

$$\det[\mathbf{Q} + \rho(\mathbf{R} + \mathbf{R}^T) + \rho^2 \mathbf{T}] = 0. \tag{4.7}$$

Since Eq. (4.7) has no real roots [1], the roots of Eq. (4.7) with positive imaginary parts are denoted by ρ_α and the corresponding eigenvectors of Eq. (4.6) by $\mathbf{a}_\alpha$ (subscript α takes values 1, 2, 3, 4). General solution to Eqs. (4.1)–(4.2) can be presented in the following form:

$$\mathbf{V} = \mathbf{A}\mathbf{f}(z) + \overline{\mathbf{A}\mathbf{f}}(\overline{z}), \tag{4.8}$$

where
$\mathbf{A} = [\mathbf{a}_1, \mathbf{a}_2, \mathbf{a}_3, \mathbf{a}_4]$ is a matrix of eigenvectors,

$$z_\alpha = x_1 + \rho_\alpha x_3,$$

$\mathbf{f}(z) = \{f_1(z_1), f_2(z_2), f_3(z_3), f_4(z_4)\}^T$ - an arbitrary vector-function that must be found.

Introducing the vector
$$\mathbf{t} = \{\sigma_{13}, \sigma_{23}, \sigma_{33}, D_3\}^T$$

and using Eqs. (4.2) and (4.8), this vector can be represented in the form

$$\mathbf{t} = \mathbf{B}\mathbf{f}'(z) + \overline{\mathbf{B}\mathbf{f}}'(\overline{z}), \tag{4.9}$$

where 4×4 matrix $\mathbf{B}$ and vector $\mathbf{f}'(z)$ are defined as following:

$$B_{KL} = (E_{2KJ1} + p_L E_{2KJ2})A_{JL},$$

$$\mathbf{f}'(z) = \left[\frac{d f_1(z_1)}{d z_1}, \frac{d f_2(z_2)}{d z_2}, \frac{d f_3(z_3)}{d z_3}, \frac{d f_4(z_4)}{d z_4}, \right]^T.$$

4.3 Fully Electrically Permeable Cracks

4.3.1 Solution to the Problem Within the Classical Model

Consider a periodic set of cracks located at the interface of two dissimilar piezo-electric half-spaces. The characteristics of the materials are assigned by their elastic moduli $C_{ijkl}^{(m)}$, piezoelectric constants $e_{ikl}^{(m)}$ and dielectric constants $\varepsilon_{il}^{(m)}$. Superscript "1" ("2") corresponds to the "upper" ("lower") half-space.

Uniformly distributed tensile (σ) and shear (τ) stresses are assumed to act at infinity and a uniformly distributed electric displacement d is assigned as well.

The coordinates of crack tips are denoted by c and b in the fundamental strip [2], and the period is assumed to be equal to h (Fig. 4.1). Additionally, the union of the open crack faces is denoted by M, the bonded parts of the interface - by U.

The open parts of the cracks are assumed to be load-free, and the crack faces are assumed to be electrically permeable. In such case, the continuity and boundary conditions for the electromechanical fields disturbed by the cracks can be written in the form

$$\langle \mathbf{V} \rangle = 0, \quad \langle \mathbf{t} \rangle = 0, \quad x_1 \in U, \tag{4.10}$$

$$\sigma_{13}^{\pm} = 0, \quad \sigma_{33}^{\pm} = 0, \quad \langle \varphi \rangle = 0, \quad \langle D_3 \rangle = 0, \quad x_1 \in M, \tag{4.11}$$

where angle brackets $\langle \rangle$ mean a jump of some function passing through the interface of materials.

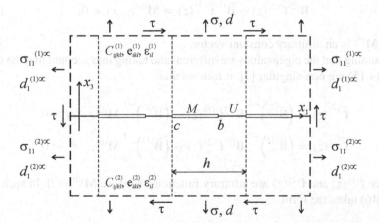

Fig. 4.1 Periodic set of fully permeable interface cracks in the piezoelectric bimaterial

General analytical solution. According to Eqs. (4.8) and (4.9), the solution to the problem (4.1) and (4.2) for each of the half-planes can be written in the form

$$\mathbf{V}^{(j)}(x_1, x_3) = \mathbf{A}^{(j)}\mathbf{f}^{(j)}(z) + \overline{\mathbf{A}}^{(j)}\overline{\mathbf{f}}^{(j)}(\overline{z}), \tag{4.12}$$

$$\mathbf{t}^{(j)}(x_1, x_3) = \mathbf{B}^{(j)}\mathbf{f}'^{(j)}(z) + \overline{\mathbf{B}}^{(j)}\overline{\mathbf{f}}'^{(j)}(\overline{z}). \tag{4.13}$$

The vector-functions $\mathbf{f}^{(1)}(z)$ and $\mathbf{f}^{(2)}(z)$ are analytical in the upper and lower half-planes, respectively.

Equation (4.13) together with the boundary conditions (4.10) lead to the following equality:

$$\mathbf{B}^{(1)}\mathbf{f}'^{(1)}(x_1) - \overline{\mathbf{B}}^{(2)}\overline{\mathbf{f}}'^{(2)}(x_1) = \mathbf{B}^{(2)}\mathbf{f}'^{(2)}(x_1) - \overline{\mathbf{B}}^{(1)}\overline{\mathbf{f}}'^{(1)}(x_1), \quad -\infty < x_1 < \infty. \tag{4.14}$$

In the last formula, the expression on the right hand side is a limit value of the analytical function in the upper half-plane, and on the left - in the lower one. Equation (4.14) indicates that both functions can be analytically extended to the whole plane as some vector-function $\mathbf{M}(z)$ which is analytical throughout the plane. Taking into account that the stresses and the electric displacement are limited at infinity, one gets from Eq. (4.13) that $\mathbf{M}(z)\big|_{z\to\pm i\infty} = \mathbf{M}^{(0)} = const$. According to Liouville's theorem this means that equality $\mathbf{M}(z) = \mathbf{M}^{(0)}$ holds in the whole plane. Therefore, it follows from Eq. (4.14) that

$$\mathbf{B}^{(1)}\mathbf{f}'^{(1)}(z) - \overline{\mathbf{B}}^{(2)}\overline{\mathbf{f}}'^{(2)}(z) = \mathbf{M}^{(0)}, \quad x_3 > 0,$$

$$\mathbf{B}^{(2)}\mathbf{f}'^{(2)}(z) - \overline{\mathbf{B}}^{(1)}\overline{\mathbf{f}}'^{(1)}(z) = \mathbf{M}^{(0)}, \quad x_3 < 0, \tag{4.15}$$

where $\mathbf{M}^{(0)}$ is an arbitrary constant vector.

Assuming that the eigenvalues are different and taking into account that the matrices in (4.15) are non-singular [1], it follows that

$$\overline{\mathbf{f}}'^{(2)}(z) = \left(\overline{\mathbf{B}}^{(2)}\right)^{-1}\mathbf{B}^{(1)}\mathbf{f}'^{(1)}(z) - \left(\overline{\mathbf{B}}^{(2)}\right)^{-1}\mathbf{M}^{(0)}, \quad x_3 > 0,$$

$$\overline{\mathbf{f}}'^{(1)}(z) = \left(\overline{\mathbf{B}}^{(1)}\right)^{-1}\mathbf{B}^{(2)}\mathbf{f}'^{(2)}(z) - \left(\overline{\mathbf{B}}^{(1)}\right)^{-1}\mathbf{M}^{(0)}, \quad x_3 < 0. \tag{4.16}$$

Since $\mathbf{f}^{(1)}(z)$ and $\mathbf{f}^{(2)}(z)$ are arbitrary functions, suppose $\mathbf{M}^{(0)} = 0$. In such case, Eq. (4.16) takes the form

$$\overline{\mathbf{f}}'^{(2)}(z) = \left(\overline{\mathbf{B}}^{(2)}\right)^{-1}\mathbf{B}^{(1)}\mathbf{f}'^{(1)}(z), \quad x_3 > 0,$$

$$\overline{\mathbf{f}}'^{(1)}(z) = \left(\overline{\mathbf{B}}^{(1)}\right)^{-1}\mathbf{B}^{(2)}\mathbf{f}'^{(2)}(z), \quad x_3 < 0. \tag{4.17}$$

Using Eqs. (4.12) and (4.17), the derivative of the displacement jump through the interface of materials can be written as

$$\langle \mathbf{V}'(x_1) \rangle = \mathbf{D}\mathbf{f}'^{(1)}(x_1) + \overline{\mathbf{D}}\overline{\mathbf{f}}'^{(1)}(x_1),\tag{4.18}$$

where

$$\mathbf{D} = \mathbf{A}^{(1)} - \overline{\mathbf{A}}^{(2)}\left(\overline{\mathbf{B}}^{(2)}\right)^{-1}\mathbf{B}^{(1)}.$$

According to Eq. (4.13), the vector $\mathbf{t}^{(1)}$ at the interface of materials can be written as

$$\mathbf{t}^{(1)}(x_1, 0) = \mathbf{B}^{(1)}\mathbf{f}'^{(1)}(x_1) + \overline{\mathbf{B}}^{(1)}\overline{\mathbf{f}}'^{(1)}(x_1).\tag{4.19}$$

Introducing the vector-function $\mathbf{W}(z)$, which is analytical in the whole plane except for the region of cracks, by the formula

$$\mathbf{W}(z) = \begin{cases} \mathbf{D}\mathbf{N}(z), & x_3 > 0, \\ -\overline{\mathbf{D}}\overline{\mathbf{N}}(z), & x_3 < 0, \end{cases}\tag{4.20}$$

where $\mathbf{N}(z) = \{f_1'^{(1)}(z), f_2'^{(1)}(z), f_3'^{(1)}(z), f_4'^{(1)}(z)\}$, the following expressions are derived for the jump of the displacements and the electric potential, and for the components of the stress tensor and the vector of the electric displacement at the interface of materials:

$$\langle \mathbf{V}'(x_1) \rangle = \mathbf{W}^+(x_1) - \mathbf{W}^-(x_1),\tag{4.21}$$

$$\mathbf{t}^{(1)}(x_1) = \mathbf{G}\mathbf{W}^+(x_1) - \overline{\mathbf{G}}\mathbf{W}^-(x_1),\tag{4.22}$$

where $\mathbf{G} = \mathbf{B}^{(1)}\mathbf{D}^{-1}$.

The components of a composite are assumed to be materials of class 6 mm [3] polarized in the direction of the x_3-axis. In that case, the plane deformation in the plane x_1x_3 takes place and the matrix $\mathbf{G}$ has the following structure after crossing out the second row and the second column:

$$\mathbf{G} = \begin{bmatrix} ig_{11} & g_{13} & g_{14} \\ g_{31} & ig_{33} & ig_{34} \\ g_{41} & ig_{43} & ig_{44} \end{bmatrix}.$$

Every g_{ij} $(i, j = 1, 3, 4)$ is real and defined by the electromechanical constants of the piezoelectric composite.

Introducing a new function

$$F_j(z) = t_j\left(W_1(z) + is_j W_2(z)\right) + \frac{\sigma_0}{1 + \gamma_j},\tag{4.23}$$

which is analytical in the whole plane except for M, and combining relations (4.21) and (4.22), the following expressions for the stresses and the displacement jump at the interface of materials are obtained:

$$\sigma_{33}^{(1)}(x_1, 0) - im_j\sigma_{13}^{(1)}(x_1, 0) = F_j^+(x_1) + \gamma_j F_j^-(x_1), \qquad (4.24)$$

$$\langle u_1'(x_1, 0)\rangle + is_j\langle u_3'(x_1, 0)\rangle = \left(F_j^+(x_1) - F_j^-(x_1)\right)/t_j, \qquad (4.25)$$

where

$$m_{1,2} = \pm\sqrt{\frac{g_{33}g_{31}}{g_{11}g_{13}}}, \quad s_j = \frac{g_{33} + m_jg_{13}}{g_{31} - m_jg_{11}}, \quad t_j = g_{31} + m_jg_{11},$$

$$\gamma_j = \frac{-g_{31} + m_jg_{11}}{m_jg_{11} + g_{31}}, \quad \sigma_0 = -\frac{g_{34}}{g_{33}g_{44} - g_{34}g_{43}}(g_{43}\sigma - g_{33}d).$$

Based on Eq. (4.24), the behaviour of the function $F_j(z)$ at infinity is

$$F_j(z)\Big|_{z\to\pm i\infty} = \frac{\sigma - im_j\tau}{1 + \gamma_j}, \quad j = 1, 2. \qquad (4.26)$$

Using relations (4.24)–(4.25), it is possible to formulate a wide range of problems of electroelasticity for a piezoelectric bimaterial with cracks.

Satisfying boundary conditions (4.11) by means of Eq. (4.24), the following boundary value problem is obtained:

$$F_j^+(x_1) + \gamma_j F_j^-(x_1) = 0, \quad x_1 \in M. \qquad (4.27)$$

It is sufficient to consider (4.27) only for $j = 1$ to find all necessary characteristics. Therefore, the subscript j is further omitted assuming that the solution is constructed for $j = 1$. Thus, the problem is reduced to determination of only one function $F(z)$, which is analytical in the entire plane except for M, where Eq. (4.27) must be satisfied.

Problem (4.27) is a periodic homogeneous Riemann boundary value problem. According to [2], its solution will be searched in the form

$$F(z) = \frac{C_1\cos\frac{\pi\left(z - \frac{c+b}{2}\right)}{h} + C_2\sin\frac{\pi\left(z - \frac{c+b}{2}\right)}{h}}{\sqrt{\Xi(z-c)\Xi(z-b)}}\left[\frac{\Xi(z-c)}{\Xi(z-b)}\right]^{i\varepsilon}, \qquad (4.28)$$

where C_1, C_2 are arbitrary complex constants which will be determined from the behavior of the function $F(z)$ at infinity ($z \to \pm i\infty$).

Consider the behavior of the function $F(z)$ at infinity. Since the argument of the trigonometric functions in Eq. (4.28) is complex and it is necessary to investigate the behavior of $F(z)$ at infinity along the imaginary axis (x_3-axis), the functions dependent on x_3 should be considered separately.

Let us transform Eq. (4.28) to get rid of the complex argument of trigonometric functions:

$$
F(z) = \left[\frac{\sin \frac{\pi(x_1 - c)}{h} + i\cos \frac{\pi(x_1 - c)}{h} \tanh \frac{\pi x_3}{h}}{\sin \frac{\pi(x_1 - b)}{h} + i\cos \frac{\pi(x_1 - b)}{h} \tanh \frac{\pi x_3}{h}} \right]^{i\varepsilon} \times
$$

$$
\times \frac{\left(C_1 \left[\cos \frac{\pi(x_1 - \frac{c+b}{2})}{h} - i\sin \frac{\pi(x_1 - \frac{c+b}{2})}{h} \tanh \frac{\pi x_3}{h} \right] + \\ +C_2 \left[\sin \frac{\pi(x_1 - \frac{c+b}{2})}{h} + i\cos \frac{\pi(x_1 - \frac{c+b}{2})}{h} \tanh \frac{\pi x_3}{h} \right] \right)}{\sqrt{\left[\sin \frac{\pi(x_1 - c)}{h} + i\cos \frac{\pi(x_1 - c)}{h} \tanh \frac{\pi x_3}{h} \right] \left[\sin \frac{\pi(x_1 - b)}{h} + i\cos \frac{\pi(x_1 - b)}{h} \tanh \frac{\pi x_3}{h} \right]}}.
$$

Tending x_3 to $+\infty$ leads to

$$
F(z)\Big|_{z \to +i\infty} = \left[\frac{\sin \frac{\pi(x_1 - c)}{h} + i\cos \frac{\pi(x_1 - c)}{h}}{\sin \frac{\pi(x_1 - b)}{h} + i\cos \frac{\pi(x_1 - b)}{h}} \right]^{i\varepsilon} \times
$$

$$
\times \frac{C_1 \left[\cos \frac{\pi(x_1 - \frac{c+b}{2})}{h} - i\sin \frac{\pi(x_1 - \frac{c+b}{2})}{h} \right] + C_2 \left[\sin \frac{\pi(x_1 - \frac{c+b}{2})}{h} + i\cos \frac{\pi(x_1 - \frac{c+b}{2})}{h} \right]}{\sqrt{\left[\sin \frac{\pi(x_1 - c)}{h} + i\cos \frac{\pi(x_1 - c)}{h} \right] \left[\sin \frac{\pi(x_1 - b)}{h} + i\cos \frac{\pi(x_1 - b)}{h} \right]}}
$$

or, after transformations,

$$
F(z)\Big|_{z \to +i\infty} = \frac{(C_1 + iC_2)e^{-i\frac{\pi(x_1 - \frac{c+b}{2})}{h}}}{ie^{-i\frac{\pi(2x_1 - c - b)}{2h}}} \frac{e^{-i\frac{\pi(x_1 - c)}{h}}}{e^{-i\frac{\pi(x_1 - b)}{h}}}.
$$

The variable x_1 is mutually reduced in the numerator and denominator, therefore the following remains:

$$
F(z)\Big|_{z \to +i\infty} = (C_2 - iC_1)e^{\frac{\pi\varepsilon(b-c)}{h}}.
$$

Similarly for x_3 tending to $-\infty$, it follows that

$$
F(z)\Big|_{z \to -i\infty} = (C_2 + iC_1)e^{-\frac{\pi\varepsilon(b-c)}{h}}.
$$

Using formula (4.26) and the last two expressions, a system of linear algebraic equations is obtained

$$\begin{cases} (C_2 - iC_1)e^{\frac{\pi\varepsilon(b-c)}{h}} = \tilde{\sigma} - im\tilde{\tau}, \\ (C_2 + iC_1)e^{-\frac{\pi\varepsilon(b-c)}{h}} = \tilde{\sigma} - im\tilde{\tau}. \end{cases}$$

Here $\tilde{\sigma} - im\tilde{\tau} = \dfrac{\sigma - im\tau}{1 + \gamma}$.

Solution of this system gives

$$C_1 = -\sin\left[i\frac{\pi\varepsilon(b-c)}{h}\right](\tilde{\sigma} - im\tilde{\tau}), \quad C_2 = \cos\left[i\frac{\pi\varepsilon(b-c)}{h}\right](\tilde{\sigma} - im\tilde{\tau}).$$

$$(4.29)$$

Thus

$$F(z) = \frac{\varXi\left(z - \frac{c+b}{2} - i\varepsilon(b-c)\right)}{\sqrt{\varXi(z-c)\varXi(z-b)}}\left[\frac{\varXi(z-c)}{\varXi(z-b)}\right]^{i\varepsilon}\frac{\sigma - im\tau}{1 + \gamma}. \qquad (4.30)$$

Determination of SIFs. Using relations (4.24) and (4.25) and the solution (4.30) of the BVP, the following expressions are obtained for the stresses and derivative of the displacement jump at the interface of materials:

$$\sigma_{33}^{(1)}(x_1, 0) - im\sigma_{13}^{(1)}(x_1, 0) =$$

$$= \frac{\varXi\left(x_1 - \frac{c+b}{2} - i\varepsilon(b-c)\right)}{\sqrt{\varXi(x_1-c)\varXi(x_1-b)}}\left[\frac{\varXi(x_1-c)}{\varXi(x_1-b)}\right]^{i\varepsilon}(\sigma - im\tau), \quad x_1 \in U, \qquad (4.31)$$

$$\langle u_3'(x_1)\rangle = \frac{1}{s_1 t_1}\, \mathrm{Im}\left[\frac{\varXi\left(x_1 - \frac{c+b}{2} - i\varepsilon(b-c)\right)}{i\sqrt{\varXi(x_1-c)\varXi(b-x_1)}}\left[\frac{\varXi(x_1-c)}{\varXi(b-x_1)}\right]^{i\varepsilon}e^{-\pi\varepsilon}(\sigma - i\tau)\right], \, x_1 \in M. \qquad (4.32)$$

For bimaterial, the SIFs in case of the classical model differ from the generally accepted ones. They are introduced by the formula

$$K_1^{osc} + imK_2^{osc} = \lim_{x_1 \to b}\sqrt{2\pi}\,[\varXi(x_1 - b)]^{1/2+i\varepsilon}\,[\sigma_{33}(x, 0) + im\sigma_{13}(x, 0)].$$

Using Eq. (4.31), the following expression is finally obtained:

$$K_1^{osc} + imK_2^{osc} = \sqrt{2\pi}\,\varXi\left(\frac{b-c}{2} - i\varepsilon(b-c)\right)\{\varXi(b-c)\}^{i\varepsilon-1/2}(\sigma + im\tau). \qquad (4.33)$$

Analysis of the results. For investigation of the cracks interaction it is enough to consider the influence of the crack length $l = b - c$ on the SIFs for various ratios of the tensile and shear loading, and for different elastic parameters of bimaterial ε and m, which characterize the relative stiffness of the components.

Bimaterials which are taken for numerical analysis are piezoelectrics of class 6 mm, polarized in the direction x_3. Corresponding graphs for the composite consisting of

- cadmium sulfide (upper material) and barium-sodium niobate (lower material) ($m = 0.976$; $\varepsilon = 0.0158$) are denoted by I,
- CTS-19 and PZT-4 ($m = 0.9197$; $\varepsilon = 0.00934$) - by II,
- PZT-5 and PZT-4 ($m = 0.905$; $\varepsilon = 0.0102$) - by III.

Electromechanical properties of the composite components are presented in Table 4.1. Normalized (with respect to external loading only) SIFs are introduced as

$$K_i^* = K_i^{osc}/\sqrt{\sigma^2 + (m\tau)^2}.$$

Dependence of the normalized SIF K_2^{osc} on the crack length is presented in Fig. 4.2. Dependence of the normalized SIF K_1^{osc} on the inclination angle of the external loading for the crack equal to a quarter of the period π is presented in Fig. 4.3.

As a result of the numerical analysis of the obtained solution, it is found out that the SIFs increase with decreasing the distance between the cracks (or, which is similar, increasing the crack length). Also SIFs grow with increasing parameter of the relative stiffness ε and decreasing m.

From the results of Fig. 4.2 follows that the values of the normalized K_2^{osc} distinguish for the different composite composition. Figure 4.3 shows that the normalized SIF K_1^{osc} decreases in absolute value with increasing ratio of the shear to tensile loading.

Conclusions. The exact solution of the plane problem of elasticity for a periodic set of cracks at the interface of two piezoelectric materials is obtained within the classical model. As a result of the numerical analysis of the solution, the variation of the stress intensity factors is illustrated depending on the distance between cracks and the inclination angle of the external loading. Based on the obtained results,

Table 4.1 Electromechanical properties of the composite material components

	Cadmium sulfide	Barium-sodium niobate	CTS-19	PZT-4	PZT-5
C_{11}, GPa	90.7	239	112.2	139	121
C_{33}, GPa	93.8	135	106	113	111
C_{13}, GPa	51	50	62.2	74.3	75.2
C_{44}, GPa	15	65	20	25.6	21.1
e_{31}, C/m^2	−0.24	−0.4	−3.4	−6.98	−5.4
e_{15}, C/m^2	−0.21	2.8	9.45	13.44	12.3
e_{33}, C/m^2	0.44	4.3	15.1	13.84	15.8
ε_{11}, nF/m	0.077	1.899	7.257	6.0	8.17
ε_{33}, nF/m	0.081	0.274	8.274	5.47	7.346

Fig. 4.2 Dependence of the normalized SIF K_2^{osc} on the cracked part of the interface

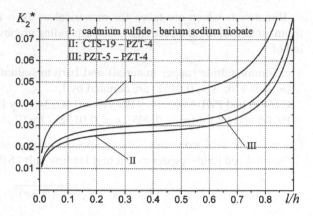

Fig. 4.3 Dependence of the normalized SIF K_1^{osc} on the angle of the external loading

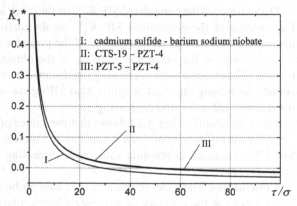

a conclusion about the crack growth resistance decrease with changing distance between the cracks can be drawn.

4.3.2 Solution to the Problem Within the Contact Model

In this subsection the same problem is considered as in the preceding one, but in view of a contact of the crack faces. Consider a periodic set of cracks located at the interface of two dissimilar piezoelectric half-planes [4, 5]. The characteristics of materials are assigned by their elastic moduli $C_{ijkl}^{(m)}$, piezoelectric constants $e_{ikl}^{(m)}$ and dielectric constants $\varepsilon_{il}^{(m)}$. Superscript "1" ("2") corresponds to the "upper" ("lower") half-plane.

Uniformly distributed tensile (σ) and shear (τ) stresses are assumed to act at infinity and a uniformly distributed electric displacement d is assigned as well.

Cracks are opened under the external electromechanical loading and a contact of faces exists in the vicinity of their tips. The coordinate of the point between the

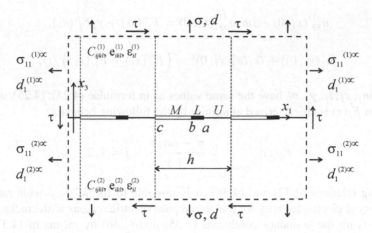

Fig. 4.4 Periodic set of fully permeable interface cracks with contact zones in the piezoelectric bimaterial

bonded interface and the open crack face in the fundamental strip [2] is denoted by c, the point of the crack faces closure is denoted by b, the point between the bonded interface and the contact zone - by a, and the period is assumed to be equal to h (Fig. 4.4). In addition, the union of the open crack faces will be denoted by M, the contact zones - by L, the bonded parts of the interface - by U. The contact zones are introduced only in the vicinity of the right crack tips. This simplifies the problem and is sound for zero and negative shear loading, since in this case the left contact zones are very small and, as noted in the previous chapters, have almost no effect on the results at the right contact zones. If the shear loading is positive, then contact zones can be considered in the vicinity of the left crack tips.

The contact zones are assumed to be smooth, the open parts of cracks - load-free, and the crack faces - electrically permeable. In that case the continuity and boundary conditions for electromechanical fields disturbed by cracks can be written in the form

$$\langle \mathbf{V} \rangle = 0, \quad \langle \mathbf{t} \rangle = 0, \quad x_1 \in U, \tag{4.34}$$

$$\sigma_{13}^{\pm} = 0, \quad \langle \sigma_{33} \rangle = 0, \quad \langle u_3 \rangle = 0, \quad \langle \varphi \rangle = 0, \quad \langle D_3 \rangle = 0, \quad x_1 \in L, \tag{4.35}$$

$$\sigma_{13}^{\pm} = 0, \quad \sigma_{33}^{\pm} = 0, \quad \langle \varphi \rangle = 0, \quad \langle D_3 \rangle = 0, \quad x_1 \in M. \tag{4.36}$$

General analytical solution. The components of a bimaterial are assumed to be piezoelectrics of class 6 mm [3] polarized in the x_3 direction. First, conduct transformations similar to (4.12)–(4.22) and introduce the function (4.23) which is analytical in the entire plane except for $L + M$. Then, combining relations (4.21) and (4.22), the following expressions for the stresses and the displacement jump at the interface of materials can be derived:

$$\sigma_{33}^{(1)}(x_1, 0) - i m_j \sigma_{13}^{(1)}(x_1, 0) = F_j^+(x_1) + \gamma_j F_j^-(x_1), \qquad (4.37)$$

$$\langle u_1'(x_1, 0)\rangle + i s_j \langle u_3'(x_1, 0)\rangle = \left(F_j^+(x_1) - F_j^-(x_1)\right)/t_j, \qquad (4.38)$$

where $m_j, s_j, t_j, \gamma_j, \sigma_0$ have the same values as in formulae (4.24), (4.25), and the function $F_j(z)$ is characterized at infinity by the following behavior:

$$F_j(z)\bigg|_{z \to \pm i\infty} = \frac{\sigma - i m_j \tau}{1 + \gamma_j}, \quad j = 1, 2. \qquad (4.39)$$

Using relations (4.37) and (4.38), it is possible to formulate a wide range of problems of electroelasticity for bimaterial piezoelectric regions with cracks.

Satisfying the boundary conditions (4.35) and (4.36) by means of (4.37) and (4.38), the following boundary value problems is obtained:

$$\begin{cases} F_j^+(x_1) + \gamma F_j^-(x_1) = 0, & x_1 \in M, \\ \operatorname{Im} F_j^\pm(x_1) = 0, & x_1 \in L. \end{cases} \qquad (4.40)$$

It is sufficient to consider the problem (4.40) only for $j = 1$ to find all necessary characteristics. Therefore, the subscript j is omitted further assuming that the solution is constructed for $j = 1$.

Thus, the problem is reduced to determination of only one function $F(z)$, which is analytical in the entire plane except for $L \cup M$, where equalities (4.40) must be satisfied.

Problem (4.40) is a periodic homogeneous Dirichlet-Riemann boundary value problem. Its solution can be presented similarly to Chap. 2 in the form

$$F(z) = \frac{e^{i\varphi(z)}}{\sqrt{\Xi(z-c)}} \left(\frac{P(z)}{\sqrt{\Xi(z-b)}} + i \frac{Q(z)}{\sqrt{\Xi(z-a)}} \right), \qquad (4.41)$$

where

$$\varphi(z) = 2\varepsilon \ln \left(\frac{\sqrt{\Xi(a-b)\Xi(z-c)}}{\sqrt{\Xi(a-c)\Xi(z-b)} + \sqrt{\Xi(b-c)\Xi(z-a)}} \right),$$

$P(z) = C_1 \cos[\pi(z - a_*)/h] + C_2 \sin[\pi(z - a_*)/h], \quad a_* = (c + b)/2,$

$Q(z) = D_1 \cos[\pi(z - b_*)/h] + D_2 \sin[\pi(z - b_*)/h], \quad b_* = (c + a)/2,$

C_1, C_2, D_1, D_2 are arbitrary real constants.

Real constants C_1, C_2, D_1, D_2 are determined from the behavior of the function $F(z)$ at infinity ($z \to \pm i\infty$). Conducting an analysis similarly to previous chapters, it follows that

$$C_1 = \sinh \chi (-m\tau \cos \zeta + \sigma \sin \zeta)/(1+\gamma),$$
$$C_2 = \cosh \chi (\sigma \cos \zeta + m\tau \sin \zeta)/(1+\gamma),$$
$$D_1 = -\sinh \chi (\sigma \cos \zeta + m\tau \sin \zeta)/(1+\gamma), \qquad (4.42)$$
$$D_2 = \cosh \chi (-m\tau \cos \zeta + \sigma \sin \zeta)/(1+\gamma),$$

where

$$\chi = 2\varepsilon \arctan \frac{\sqrt{\sin \frac{\pi(a-c)}{h}} \sin \frac{\pi(b-c)}{2h} + \sqrt{\sin \frac{\pi(b-c)}{h}} \sin \frac{\pi(a-c)}{2h}}{\sqrt{\sin \frac{\pi(a-c)}{h}} \cos \frac{\pi(b-c)}{2h} + \sqrt{\sin \frac{\pi(b-c)}{h}} \cos \frac{\pi(a-c)}{2h}},$$

$$\zeta = \varepsilon \ln \frac{\sin \frac{\pi(a+b-2c)}{2h} + \sqrt{\sin \frac{\pi(a-c)}{h}} \sin \frac{\pi(b-c)}{h}}{\sin \frac{\pi(a-b)}{2h}}. \qquad (4.43)$$

Derivation of the Classical "Oscillating" Solution as a Particular Case of the "Contact" Model Solution

The solution within the contact model can be analytically reduced to the oscillating one assuming that the contact zone length tends to zero. After transformations similar to Sects. 2.1.3 and 3.1.2, it follows that

$$F(z)\Big|_{b \to a} = \frac{\Xi\left(z - \frac{c+a}{2} - i\varepsilon(a-c)\right)}{\sqrt{\Xi(z-c)\Xi(z-a)}} \left[\frac{\Xi(z-c)}{\Xi(z-a)}\right]^{i\varepsilon} \frac{\sigma - im\tau}{1+\gamma}. \qquad (4.44)$$

The obtained expression for the function $F(z)$ coincides with corresponding expression (4.30) for the classical model.

Determination of the contact zones and intensity factors. Using the relations (4.37) and (4.38) and the solution (4.41), the following expressions for the stresses and the derivative of displacement jump at the interface of materials can be obtained:

$$\sigma_{33}^{(1)}(x_1, 0) = \frac{2ge^{\pi\varepsilon}}{\sqrt{\Xi(x_1 - c)}} \times$$

$$\times \left(\frac{P(x_1)}{\sqrt{\Xi(x_1 - b)}} \cosh(\tilde{\varphi}(x_1) - \pi\varepsilon) + \frac{Q(x_1)}{\sqrt{\Xi(a - x_1)}} \sinh(\tilde{\varphi}(x_1) - \pi\varepsilon)\right), \quad x_1 \in L,$$

$$\qquad (4.45)$$

$$\sigma_{33}^{(1)}(x_1, 0) - im\sigma_{13}^{(1)}(x_1, 0) =$$

$$= \frac{g(1+\gamma)e^{i\varphi(x_1)}}{\sqrt{\Xi(x_1 - c)}} \left(\frac{P(x_1)}{\sqrt{\Xi(x_1 - b)}} + \frac{iQ(x_1)}{\sqrt{\Xi(x_1 - a)}}\right), \quad x_1 \in U, \qquad (4.46)$$

$$\langle u_3'(x_1) \rangle = \frac{-2\cosh(\pi\varepsilon)}{s_1 t_1 \sqrt{\Xi(x_1 - c)}} \times$$

$$\times \left(\frac{P(x_1)}{\sqrt{\Xi(b - x_1)}} \cos\varphi^*(x_1) - \frac{Q(x_1)}{\sqrt{\Xi(a - x_1)}} \sin\varphi^*(x_1) \right), \quad x_1 \in M, \quad (4.47)$$

where

$$\tilde{\varphi}(x_1) = 2\varepsilon \arctan\sqrt{\frac{\Xi(b - c)\Xi(a - x_1)}{\Xi(a - c)\Xi(x_1 - b))}}, \quad x_1 \in L,$$

$$\varphi^*(x_1) = 2\varepsilon \ln\frac{\sqrt{\Xi(a - b)\Xi(x_1 - c)}}{\sqrt{\Xi(a - c)\Xi(b - x_1)} + \sqrt{\Xi(b - c)\Xi(a - x_1)}}, \quad x_1 \in M.$$

Equations (4.45)–(4.47) can be used for any position of the point b, but the above obtained solution will be physically sound under the condition that the stresses are compressive in the contact zone, the crack faces do not interpenetrate each other and close smoothly in the point b. These conditions can be written mathematically as

$$\langle u_3'(b) \rangle = 0; \quad \sigma_{33}^{(1)}(x_1, 0) \leq 0, \quad x_1 \in L; \quad \langle u_3(x_1) \rangle \geq 0, \quad x_1 \in M. \quad (4.48)$$

Using condition (4.48$_1$) and expanding Eq. (4.47) into Taylor series in the vicinity of the point b, the transcendental equation is obtained to determine the contact zone length

$$P(b) = 0. \quad (4.49)$$

The correct selection of the root is provided by satisfaction of the conditions (4.48$_2$) and (4.48$_3$). Equation (4.49) can be rewritten as follows:

$$\tan\frac{\pi(b - c)}{2h} = -\frac{-m\tau\cos\zeta + \sigma\sin\zeta}{\sigma\cos\zeta + m\tau\sin\zeta}\tanh\chi. \quad (4.50)$$

The SIFs at the crack tip a are defined by the formula

$$K_1 - iK_2 = \lim_{x_1 \to a+0}(\sigma_{33}(x_1, 0) - i\sigma_{13}(x_1, 0))\sqrt{\Xi(x_1 - a)}.$$

Usage of Eq. (4.46) leads to the following formulae:

$$K_1 = 0,$$

$$K_2 = -\frac{1 + \gamma}{m\sqrt{\Xi(a - c)}}\left(D_1\cos\frac{\pi(a - c)}{2h} + D_2\sin\frac{\pi(a - c)}{2h}\right) = \quad (4.51)$$

$$= \frac{(\sigma\cos\zeta + m\tau\sin\zeta)\sinh\chi\cos\frac{\pi(a-c)}{2h} + (m\tau\cos\zeta - \sigma\sin\zeta)\cosh\chi\sin\frac{\pi(a-c)}{2h}}{m\sqrt{\Xi(a - c)}}.$$

Analysis of the results. For the investigation of the periodic set of interface cracks it is enough to consider the dependence of the relative contact zone length $\lambda = (a - b)/l$ and the SIF at the crack tip a on the crack length l and the inclination angle β of the vector of resulting loading $p = \sqrt{\sigma^2 + \tau^2}$ to the x_3-axis (tan $\beta = \tau/\sigma$). The parameter β varies in the range $-\pi/2 \le \beta \le \pi/2$. Positive (negative) values of the parameter ε mean that the "lower" ("upper") material is more stiff than the "upper" ("lower") one.

For numerical analysis consider bimaterials consisting of piezoelectrics of class 6 mm, polarized in the x_3 direction. Corresponding graphics for the composite consisting of cadmium sulfide (upper material) and barium-sodium niobate (lower material) ($m = 0.976$; $\varepsilon = 0.0158$) are denoted by I, PZT-5 and PZT-4 ($m = 0.905$; $\varepsilon = 0.0102$) - by II.

Figure 4.5 shows the dependence of the relative contact zone length λ on the distance between cracks under the remote combined ($\sigma = 1$ MPa, $\tau = -5$ MPa) loading. The corresponding results for the dimensionless SIF $K^* = K_2/(p\sqrt{l})$ under the same loading are shown in Fig. 4.6. As it is expected, a decrease of the distance between the cracks significantly influences the contact zone length and the SIF only for l/h close to 1.

The results for the relative contact zone length under larger shear loading ($\sigma = 1$ MPa, $\tau = -10$ MPa) than in Fig. 4.5 are presented in Fig. 4.7. A comparison with Fig. 4.5 shows that increasing the value of the shear loading leads to the contact zone increase since the upper material is less stiff than the lower one. It should be noted that in such direction of the shear loading application, the contact zone formed in the vicinity of the investigated crack tip is larger than in the vicinity of the left tip.

Figure 4.8 shows a comparison of the normal stress field (in MPa) on the extension of the crack along the interface of materials, obtained within the oscillating and contact models under the combined loading. The composite consists of cadmium sulfide (upper material) and barium sodium niobate (lower material). The ratio of the tensile loading to the shear one is equal to -10. The crack length is half the period and the relative contact zone length is equal to 8.24595×10^{-4}, $x_1' = (x_1 - a)/l$. It

Fig. 4.5 Dependence of the relative contact zone length λ on the distance between cracks under combined loading

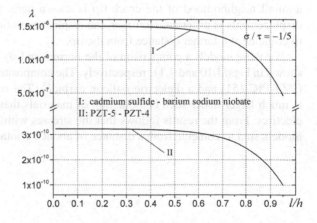

Fig. 4.6 Dependence of the dimensionless SIF on the distance between cracks under combined loading

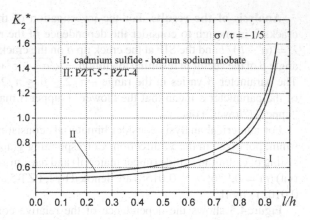

Fig. 4.7 Dependence of the relative contact zone length λ on the distance between cracks under larger shear loading than in Fig. 4.5

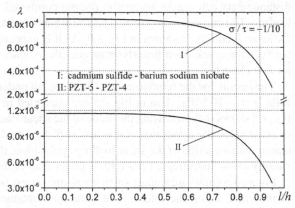

is seen from the results that stresses obtained from the oscillating and contact models start to coincide from a distance of about 6 contact zone lengths.

A comparison of the normal displacement jump (in mm) is shown in Fig. 4.9 within the oscillating and contact models when approaching the right crack tip. Only a small neighborhood of the crack tip is shown here, where the behavior of the displacements is different. However, the displacement fields for both models begin to coincide at a farther distance from the tip.

For these models a comparison of the normal and shear stress fields (in MPa) is shown in Figs. 4.10 and 4.11, respectively. The composite consists of a piezoelectric (PZT-PIC 151) and a dielectric (silicon carbide). The relative contact zone length is much larger under such combination of materials than in the case of two piezo-electrics. From the results follows that the stresses within the classical and contact models start to coincide from a distance of about 7 contact zone lengths.

Fig. 4.8 Comparison of the normal stress fields (oscillatory vs. contact approaches) near the right crack tip under combined loading

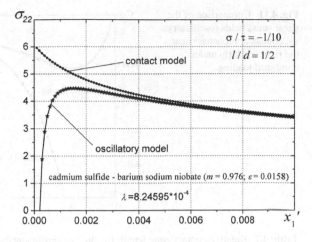

Fig. 4.9 Comparison of the normal displacement jump (oscillatory vs. contact approaches) near the right crack tip under combined loading

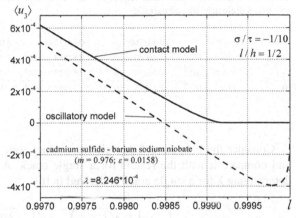

Fig. 4.10 Comparison of the normal stress fields (oscillatory vs. contact approaches) near the right crack tip under larger shear loading than in Fig. 4.8

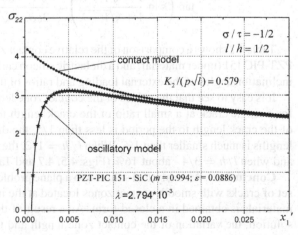

Fig. 4.11 Comparison of the shear stress fields (oscillatory vs. contact approaches) near the right crack tip under combined loading

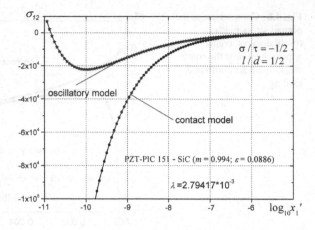

Table 4.2 Relative contact zone length for the composite consisting of a piezoelectric PZT-PIC 151 and a dielectric silicon carbide

σ/τ	Reference [6]	$l/h = 1/20$	$l/h = 1/5$	$l/h = 1/2$	$l/h = 3/4$
$-1/10$	0.141796	0.141760	0.141184	0.136212	0.119807
$-1/2$	2.86043×10^{-3}	2.86040×10^{-3}	2.85884×10^{-3}	2.79417×10^{-3}	2.38123×10^{-3}
-1	7.56280×10^{-5}	7.56279×10^{-5}	7.55927×10^{-5}	7.39056×10^{-5}	6.29257×10^{-5}
0	1.11215×10^{-8}	1.11215×10^{-8}	1.11163×10^{-8}	1.08683×10^{-8}	9.25345×10^{-9}

Consider the behavior of the relative contact zone length for a small ratio l/h and compare it with the results for a single crack. A transcendental equation for determining λ for a single crack is derived in [6]:

$$\tan\left(\varepsilon \ln \frac{1 - \sqrt{1 - \lambda}}{1 + \sqrt{1 - \lambda}}\right) = \frac{\sqrt{1 - \lambda}\sigma + 2\varepsilon m\tau}{2\varepsilon\sigma - \sqrt{1 - \lambda}m\tau}. \tag{4.52}$$

Table 4.2 shows a comparison of the relative contact zone length for a piezoelectric PZT-PIC 151 (upper material) and a dielectric silicon carbide (lower material) at some inclination angles of the external loading and ratios of the crack length to the period.

It is easy to see that the values of the contact zone length coincide with the results for a single crack at a small ratio of the crack length to the period. When the ratio of the crack length to the period is less than $1/5$, the difference in the contact zone lengths is much smaller than 1%. When $l/h = 1/2$, the difference is on average 3%, and when $l/h = 3/4$ - about 16% (Figs. 4.5, 4.7 and Table 4.2).

Conclusions. An analytical solution to a plane problem of elasticity for a periodic set of cracks with smooth contact zones located at the interface of two piezoelectric materials is obtained in a closed form. As a result of the numerical analysis of this solution, the variation of the contact zone length and the stress intensity factors is considered depending on the distance between cracks and the inclination angle of the external loading. Based on the obtained results, a conclusion about the crack growth resistance decrease with decreasing distance between the cracks is drawn. The results for the case of a small crack length (with respect to the period) are compared with the results for a single crack. A solution for the periodic set of electrically permeable cracks within the oscillating model is derived as a particular case. A comparison of the normal and shear stress fields within the oscillating and contact models is considered on the extension of the crack along the interface of materials.

4.3.3 Fracture Behavior of Periodically Bonded Piezoelectric Interface Under Compressive-Shear Loading

Most of the interface delamination studies were performed under purely tensile or tensile-shear loading which is the most dangerous with respect to the fracture of the composite. However, dielectric and piezoelectric materials with interface defects are often under the action of compressive or compressive-shear loading (Fig. 4.12). Being completely closed under pure compression, the interface partially opens upon addition of the shear field.

To explore a piezoelectric composite whose components are bonded periodically [7], we investigate the dependence of the relative contact zone length $\lambda = (a - b)/l$ and the SIF K_2 at the crack tip a on

- the ratio of the length $l = a - c$ (the unbonded part of the interface or crack) to the period h.
- the angle β between the resulting vector p of the applied compressive-shear $\sigma - \tau$ loading and the axis x_3: $\tan \beta = \tau/\sigma$, $p = \sqrt{\sigma^2 + \tau^2}$, $-\pi/2 \leq \beta \leq \pi/2$.

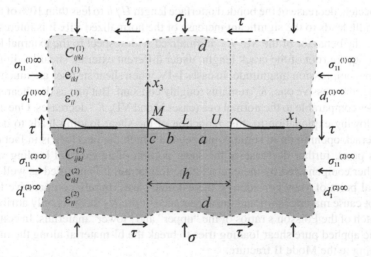

Fig. 4.12 Periodic set of fully permeable interface cracks under compressive-shear loading

For numerical calculations consider a composite of cadmium sulfide (CdS) and barium sodium niobate ($Ba_2NaNb_5O_{15}$) with $m = 0.976$; $\varepsilon = 0.0158$. The positive value of ε indicates that the "lower" material - barium sodium niobate - is more rigid. The compressive stress $\sigma = 1$ MPa as well as the shear stress corresponding to the six calculation variants I: $\tau = 1000$ MPa; II: $\tau = 50$ MPa; III: $\tau = 10$ MPa; IV: $\tau = 5$ MPa; V: $\tau = 1$ MPa; VI: $\tau = 0.5$ MPa are applied to the bi-material.

Directions of the action of the external normal and shear stresses are shown in Fig. 4.12. Under the compressive loading, the unbonded part of the interface is closed, i.e. the length of the part $M = 0$. Depending on the magnitude of the applied shear loading, partial opening of the boundary of two materials is observed.

The dependence of the relative contact area length λ on the ratio of the bonded portions of the interface to the distance between them (period length) under the remote combined loading is plotted in Fig. 4.13. An increase of the shear loading shifts the curves down from VI to I. Two limiting cases are observed: the biggest opening of the interface is under strong shear loading (case I), while when the shear loading is equal or smaller than the compressive one (cases V and VI) the opening is negligible (more than 99% of the interface region is closed). Increasing the ratio of the shear stress to the compressive one (cases II and III), the opening of the interface increases too. The relative contact zone length is decreasing as the unbonded parts of the interface (or cracks) augment, i.e. the shorter are the bonds between the materials, the smaller is the portion in contact. For example, with a crack length equal to 0.95, the relative contact zone size is almost half of that in case of a single crack. The results for a periodically bonded interface in case of large bonded parts ($(M + L)/h < 1/20$ or $U/h > 19/20$) coincide with the results obtained for a single interface crack under the same loading [8].

The influence of the bonded proportion on the normalized SIF $K^* = K_2/(p\sqrt{\pi l})$ under the different magnitudes of the shear loading is as well presented in Fig. 4.13. As expected, decrease of the bonded interface length U/h to less than 10% of the unit strip width leads to the significant increase of the normalized SIF. It is interesting to analyze the behavior of the SIF K_2, normalized with respect to the external loading and the square root of the crack length, under different external shear loading at the constant compression magnitude. In cases I–IV, when shear stresses are much bigger than the compressive one, K^* remains roughly constant. But just as the shear loading becomes comparable to the normal one (cases V and VI), K^* decreases. One can find the following explanation to this: reduction of the shear loading leads to decrease of the crack opening up to some point, beyond which the interface is in fact closed. At this point, further decrease of the shear portion of an external loading will not be further compensated by the contact zone formation. It is justified as well from a physical point of view considering the extreme cases: purely compressive loading will not cause material shift and the appearance of the K_2 factor is only attributed to mismatch of the Poisson's ratios of the "upper" and "lower" materials. In contrast to this, the applied pure shear loading tries to break the bi-material along the interface according to the Mode II fracture.

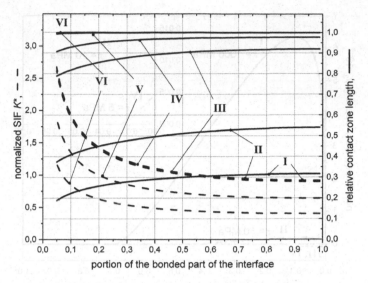

Fig. 4.13 Normalized SIF K_2 (dash lines) and contact length λ (solid lines) as functions of the ratio between the length of the bonded interface to the period (U/h) under varying magnitude of the shear loading

Figure 4.14 illustrates u_3 opening (a jump of the normal displacements) for cases I–IV. It is normalized with respect to the length of the bonded portion U of the interface.

With decrease of the shear loading compared to the compressive one, the size of the zone in contact increases, and, accordingly, the area in which the crack is open, is reduced. It is easy to observe that the maximum crack opening decreases respectively. It is interesting to note that the crack opening profile is self-similar (e.g., comparing cases I and IV).

Figure 4.15 compares the normal stresses σ_{33} along the bonded part of the interface ahead of the tip a obtained in the framework of the "oscillation" and "contact" models for dominated shear loading (case I). The bi-material consists of PZT-PIC151 and SiC.

The length of the bonded (U) and unboded ($M + L$) parts of the interface are taken equal ($l = h/2$). The contact area $\lambda = 0.300595$. It is worth to note that the normal stresses are limited near the point a in case of the contact approach. The use of the oscillatory approach leads to a singularity and change in the stress sign. In Fig. 4.15 $x_1 \rightarrow h$ corresponds to the left vertex of the open region M since we consider the periodic system: approaching the end of the period is followed by the beginning of the next identical region of the length h. Thus, the rightmost part of Fig. 4.15 corresponds to the region left of the point c (see Fig. 4.12). Since the left vertex of the opened region M is considered within the framework of the oscillation model, it can be seen that the stresses begin to coincide when approaching the left crack tip.

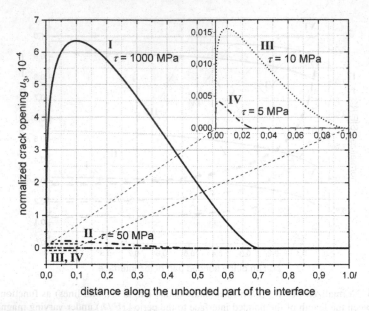

Fig. 4.14 Opening of the faces of unbonded interface and magnitude of the contact zone for the different values of external loading

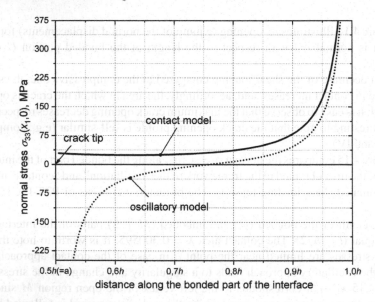

Fig. 4.15 Behavior of the normal stresses σ_{33} along the ligament ahead of the point a

Conclusions. A closed-form analytical solution of an elastic problem for a periodically bonded interface of two piezoelectric (or piezo- and dielectric) materials is obtained. It accounts for the smooth contact of the materials along the interface which can have an arbitrary size including the completely closed imperfections. As a result of the numerical analysis of the obtained solution for the different lengths of the bonded part of the interface and angles of the tension-shear loading, the change of the contact zone length and the stress intensity factor are exemplified. Based on the obtained results for the SIF, one can draw a conclusion about the degree of strength reduction of a composite material weakened due to the partially bonded interface. As a partial case of the current solution, the classical "oscillatory" solution can be easily derived. A comparison of the normal and tangent stresses along the ligament obtained in the framework of the both models was carried out.

4.4 Cracks with Finite Electric Permittivity

In this section a solution to the problem of a periodic set of cracks located at the interface of the piezoelectric bimaterial is constructed assuming that the crack filler has a finite electric permittivity. In case of a homogeneous piezoelectric material, the periodic set of cracks is positioned along a line.

4.4.1 Solution for a Homogeneous Piezoelectric Material

Statement of the problem. Consider a periodic set of cracks located along a line in a homogeneous piezoelectric half-space. The characteristics of the materials are assigned by their elastic moduli C_{ijkl}, piezoelectric constants e_{ikl} and dielectric constants ε_{il}.

Uniformly distributed tensile (σ) and shear (τ) stresses are assumed to act at infinity along with a uniformly distributed electric displacement d. The cracks are opened under the action of the external loading. The coordinates of the crack tips in a fundamental strip [2] are denoted by $-b$ and b. A period is assigned to be equal to h. The union of the cracks is denoted by M and the union of the bonded parts of the interface is denoted by U (Fig. 4.16).

Introducing the electric field inside the cracks as

$$E_a = -(\varphi^+ - \varphi^-)/(u^+ - u^-)$$

and taking into account that

$$D_3 = \varepsilon_a E_a,$$

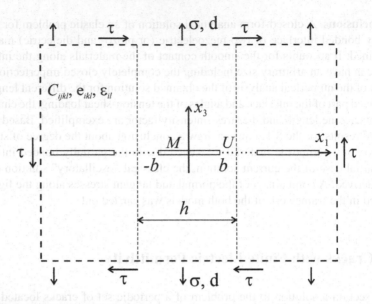

Fig. 4.16 Periodic set of semi-permeable cracks in a homogeneous piezoelectric material

the following conditions are derived in the crack areas:

$$D_3 = -\varepsilon_a \frac{\varphi^+ - \varphi^-}{u^+ - u^-}, \tag{4.53}$$

where

$\varepsilon_a = \varepsilon_0 \varepsilon_r$ is the electric permittivity of the crack filler,
$\varepsilon_0 = 8.85 \times 10^{-12}$ F/m.

The electric condition (4.53) is obtained by Parton and Kudryavtsev [3] for a thin dielectric layer and by Hao and Shen [9] for a crack.

Assuming that the open parts of cracks are load-free, the continuity and boundary conditions along the crack line (cf. Fig. 4.16) can be written in the form

$$\langle \mathbf{V}(x_1, 0) \rangle = 0, \quad \langle \mathbf{t}(x_1, 0) \rangle = 0, \quad x_1 \in U, \tag{4.54}$$

$$\sigma_{13}^{\pm}(x_1, 0) = 0, \quad \langle \sigma_{33}(x_1, 0) \rangle = 0, \quad D_3 \langle u_3(x_1, 0) \rangle = -\varepsilon_a \langle \varphi(x_1, 0) \rangle,$$
$$\langle D_3(x_1, 0) \rangle = 0, \quad x_1 \in M. \tag{4.55}$$

General analytical solution. The components of the bimaterial are assumed to be piezoceramic of class 6 mm [3] polarized in the direction of the x_3-axis. After transformations similar to (4.12)–(4.20) and combining relations (4.21) and (4.22), the following expressions for the stresses and the electric displacement are derived:

$$\sigma_{33}^{(1)}(x_1, 0) + m_{j4} D_3^{(1)}(x_1, 0) + i m_{j1} \sigma_{13}^{(1)}(x_1, 0) = F_j^+(x_1) + F_j^-(x_1), \quad (4.56)$$

where
$$F_j(z) = n_{j1} W_1(z) + i[n_{j3} W_3(z) + n_{j4} W_4(z)], \quad (4.57)$$

coefficients m_{jl}, n_{jl} ($j, l = 1, 3, 4$) depend on the electromechanical properties of the composite and are real values for the piezoceramic of class 6 mm, $\varepsilon_4 = 0$, $\gamma_4 = 1$ for ceramics of class 6 mm polarized in the direction of the x_3-axis.

The electric displacement is constant along the crack faces in case of a homogeneous material

$$D_3^+(x_1, 0) = D_3^-(x_1, 0) = D, \quad x_1 \in M. \quad (4.58)$$

Using Eqs. (4.56) and (4.58) together with conditions (4.55) at the crack faces, a periodic homogeneous Riemann problem is obtained

$$F_j^+(x_1) + F_j^-(x_1) = m_{j4} D, \quad (j = 1, 3, 4), \quad x_1 \in M. \quad (4.59)$$

By introducing a new function

$$\Phi_j(z) = F_j(z) - m_{j4} D/2, \quad (4.60)$$

Equation (4.59) takes the form

$$\Phi_j^+(x_1) + \Phi_j^-(x_1) = 0, \quad (j = 1, 3, 4), \quad x_1 \in M. \quad (4.61)$$

The conditions at infinity according to Eq. (4.56) are

$$\Phi_j(z)\Big|_{z \to \pm i\infty} = \sigma_j^* - i\tau_j^*, \quad (4.62)$$

where
$$\sigma_j^* = [\sigma + m_{j4}(d - D)]/2, \quad \tau_j^* = -m_{j1}\tau/2.$$

A solution of the periodic homogeneous Riemann problem (4.61) is searched in the form [2]

$$\Phi_j(z) = \frac{C_{1j} \cos \frac{\pi z}{h} + C_{2j} \sin \frac{\pi z}{h}}{\sqrt{\sin \frac{\pi(z-b)}{h} \sin \frac{\pi(z+b)}{h}}}. \quad (4.63)$$

Complex constants C_{1j}, C_{2j} are defined from the behavior of the function at infinity by considering $z \to \pm i\infty$.

Since the argument of the trigonometric functions is complex and it is necessary to investigate its behavior at infinity for the imaginary axis coordinate (x_3-axis), the functions dependent on x_3 should be identified separately. Transforming expression (4.63) gives

$$\Phi_j(z) = \frac{C_{1j}\left[\cos\frac{\pi x_1}{h} - i\sin\frac{\pi x_1}{h}\tanh\frac{\pi x_3}{h}\right] + C_{2j}\left[\sin\frac{\pi x_1}{h} + i\cos\frac{\pi x_1}{h}\tanh\frac{\pi x_3}{h}\right]}{\sqrt{\left[\sin\frac{\pi(x_1-b)}{h} + i\cos\frac{\pi(x_1-b)}{h}\tanh\frac{\pi x_3}{h}\right]\left[\sin\frac{\pi(x_1+b)}{h} + i\cos\frac{\pi(x_1+b)}{h}\tanh\frac{\pi x_3}{h}\right]}}.$$

Letting x_3 tend to $+\infty$ gives

$$\Phi_j(z)\bigg|_{z\to+i\infty} = \frac{C_{1j}\left[\cos\frac{\pi x_1}{h} - i\sin\frac{\pi x_1}{h}\right] + C_{2j}\left[\sin\frac{\pi x_1}{h} + i\cos\frac{\pi x_1}{h}\right]}{\sqrt{\left[\sin\frac{\pi(x_1-b)}{h} + i\cos\frac{\pi(x_1-b)}{h}\right]\left[\sin\frac{\pi(x_1+b)}{h} + i\cos\frac{\pi(x_1+b)}{h}\right]}}.$$

After transformations one gets

$$\Phi_j(z)\bigg|_{z\to+i\infty} = \frac{(C_{1j} + iC_{2j})e^{-i\frac{\pi x_1}{h}}}{ie^{-i\frac{\pi x_1}{h}}}.$$

Multipliers dependent on the variable x_1 are mutually reduced in the numerator and denominator. Therefore, the following formula is finally obtained

$$\Phi_j(z)\bigg|_{z\to+i\infty} = C_{2j} - iC_{1j}. \qquad (4.64)$$

Similarly, letting x_3 tend to $-\infty$, it follows that

$$\Phi_j(z)\bigg|_{z\to-i\infty} = C_{2j} + iC_{1j}. \qquad (4.65)$$

Using Eq. (4.62) together with Eqs. (4.64) and (4.65), a system of linear algebraic equations is obtained to find the arbitrary complex constants C_{1j} and C_{2j}. Solution of this system is

$$C_{1j} = 0, \quad C_{2j} = \sigma_j^* - i\tau_j^*.$$

Substituting the obtained values for C_{1j} and C_{2j} in Eq. (4.63) and conducting elementary transformations, the solution of the periodic homogeneous Riemann problem (4.61) takes the form

$$\Phi_j(z) = \left(\sigma_j^* - i\tau_j^*\right)\frac{\Xi(z)}{\sqrt{\Xi(z-b)\Xi(z+b)}}. \qquad (4.66)$$

Using Eq. (4.57), the following expression for the jump of the mechanical displacements and electric potential derivatives is derived:

$$n_{j1}\langle u_1'(x_1,0)\rangle + i\left(n_{j3}\langle u_3'(x_1,0)\rangle + n_{j4}\langle\varphi'(x_1,0)\rangle\right) = F_j^+(x_1) - F_j^-(x_1).$$

Since $\Phi_j^+(x_1) = -\Phi_j^-(x_1)$, $x_1 \in M$, then $F_j^+(x_1) - F_j^-(x_1) = 2\Phi_j^+(x_1)$, $x_1 \in M$ and the above formula takes the following form at the crack faces:

$$n_{j1}\langle u_1'(x_1, 0)\rangle + i \left(n_{j3}\langle u_3'(x_1, 0)\rangle + n_{j4}\langle \varphi'(x_1, 0)\rangle\right) =$$
$$= -2i \left(\sigma_j^* - i\tau_j^*\right) \frac{\Xi(x_1)}{\sqrt{\Xi(b-x_1)\Xi(b+x_1)}}, \quad x_1 \in M.$$

Integrating this expression, the following formula for the jump of the mechanical displacements and electric potential is obtained:

$$n_{j1}\langle u_1(x_1, 0)\rangle + i \left(n_{j3}\langle u_3(x_1, 0)\rangle + n_{j4}\langle \varphi(x_1, 0)\rangle\right) = 2i \left(\sigma_j^* - i\tau_j^*\right) H(x_1), \quad x_1 \in M,$$
(4.67)

where $H(x_1) = \dfrac{h}{\pi} \ln \left(\dfrac{\cos \frac{\pi x_1}{h} + \sqrt{\sin \frac{\pi(b-x_1)}{h} \sin \frac{\pi(x_1+b)}{h}}}{\cos \frac{\pi b}{h}} \right)$.

The graph of the function $H(x_1)$ for $b = 1$ and $h = \pi$ is presented in Fig. 4.17.

It should be noted that this graph is similar to the graph for a single crack in a homogeneous material described by the function $\sqrt{b^2 - x_1^2}$.

Considering the real part of Eq. (4.67) for $j = 1$, it follows that

$$\langle u_1(x_1, 0)\rangle = -\frac{m_{11}\tau}{n_{11}} H(x_1).$$

Considering the imaginary part of Eq. (4.67) for $j = 1, 4$, the following system is obtained:

$$\begin{cases} n_{13}\langle u_3(x_1, 0)\rangle + n_{14}\langle \varphi(x_1, 0)\rangle = 2\sigma_1^* H(x_1), \\ n_{43}\langle u_3(x_1, 0)\rangle + n_{44}\langle \varphi(x_1, 0)\rangle = 2\sigma_4^* H(x_1). \end{cases}$$

Solution to this system of linear algebraic equations has the form

Fig. 4.17 Graph of the function $H(x_1)$ characterizing the shape of the mechanical displacements and electric potential jump

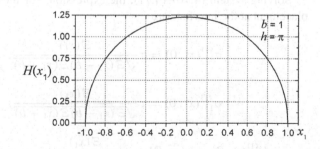

$$\langle u_3(x_1, 0) \rangle = \frac{2}{\Delta} \{ n_{44}\sigma_1^* - n_{14}\sigma_4^* \} H(x_1),$$

$$\langle \varphi(x_1, 0) \rangle = \frac{2}{\Delta} \{ n_{13}\sigma_4^* - n_{43}\sigma_1^* \} H(x_1),$$

$$\Delta = n_{13}n_{44} - n_{14}n_{43}.$$

Using relation (4.53), the following equation is obtained to determine the electric displacement D inside the cracks:

$$D = \varepsilon_a \frac{n_{43}\sigma_1^*(D) - n_{13}\sigma_4^*(D)}{n_{44}\sigma_1^*(D) - n_{14}\sigma_4^*(D)}, \tag{4.68}$$

which can be rewritten in the form

$$\eta_1^* D^2 + \eta_2^* D + \eta_3^* = 0, \tag{4.69}$$

where
$$\eta_1^* = m_{44}n_{14} - m_{14}n_{44}, \quad \eta_2^* = (s_1 n_{44} + \varepsilon_a m_{14}n_{43}) - (s_4 n_{14} + \varepsilon_a m_{44}n_{13}),$$
$$\eta_3^* = \varepsilon_a(s_4 n_{13} - s_1 n_{43}), \quad s_1 = \sigma + m_{14}d, \quad s_4 = \sigma + m_{44}d.$$
Using Eqs. (4.56), (4.60) and (4.66) for $j = 1$ and the real part of Eqs. (4.56), (4.60) and (4.66) for $j = 4$, the following system of equations is obtained to determine the stresses and electric displacement in $x_1 \in U$:

$$\sigma_{33}^{(1)}(x_1, 0) + m_{14}D_3^{(1)}(x_1, 0) + im_{11}\sigma_{13}^{(1)}(x_1, 0) =$$
$$= (\sigma + m_{14}(d - D) + im_{11}\tau)\frac{\Xi(x_1)}{\sqrt{\Xi(x_1 - b)\Xi(x_1 + b)}} + m_{14}D, \tag{4.70}$$

$$\sigma_{33}^{(1)}(x_1, 0) + m_{44}D_3^{(1)}(x_1, 0) = (\sigma + m_{44}(d - D))\frac{\Xi(x_1)}{\sqrt{\Xi(x_1 - b)\Xi(x_1 + b)}} + m_{44}D. \tag{4.71}$$

Solving system (4.70)–(4.71), the expressions for the stresses and electric displacement at the interface are

$$\sigma_{33}^{(1)}(x_1, 0) = \sigma \frac{\Xi(x_1)}{\sqrt{\Xi(x_1 - b)\Xi(x_1 + b)}}, \quad x_1 \in U, \tag{4.72}$$

$$\sigma_{13}^{(1)}(x_1, 0) = \tau \frac{\Xi(x_1)}{\sqrt{\Xi(x_1 - b)\Xi(x_1 + b)}}, \quad x_1 \in U, \tag{4.73}$$

$$D_3^{(1)}(x_1, 0) = (d - D)\frac{\Xi(x_1)}{\sqrt{\Xi(x_1 - b)\Xi(x_1 + b)}} + D, \quad x_1 \in U. \tag{4.74}$$

Introducing the intensity factors as

$$K_1 = \lim_{x_1 \to b+0} \sqrt{2\pi \, \varXi \, (x_1 - b)} \sigma_{33}^{(1)}(x_1, 0),$$

$$K_2 = \lim_{x_1 \to b+0} \sqrt{2\pi \, \varXi \, (x_1 - b)} \sigma_{13}^{(1)}(x_1, 0),$$

(4.75)

$$K_4 = \lim_{x_1 \to b+0} \sqrt{2\pi \, \varXi \, (x_1 - b)} D_3^{(1)}(x_1, 0),$$

(4.76)

and, using Eqs. (4.72)–(4.74), the IFs take the final form

$$K_1 = \sqrt{\pi \tan \frac{\pi b}{h}} \, \sigma, \quad K_2 = \sqrt{\pi \tan \frac{\pi b}{h}} \, \tau, \quad K_4 = \sqrt{\pi \tan \frac{\pi b}{h}} (d - D). \quad (4.77)$$

For the displacement and electric potential jump we obtained the following closed-form expressions:

$$\langle u_3(x_1, 0) \rangle = \{ \vartheta_{11}\sigma + \vartheta_{12}(d - D) \} H(x_1), \quad x_1 \in M, \quad (4.78)$$

$$\langle u_1(x_1, 0) \rangle = -m_{11}n_{11}^{-1}\tau H(x_1), \quad x_1 \in M, \quad (4.79)$$

$$\langle \varphi(x_1, 0) \rangle = \{ \vartheta_{21}\sigma + \vartheta_{22}(d - D) \} H(x_1), \quad x_1 \in M, \quad (4.80)$$

where

$$\vartheta_{11} = \frac{n_{44} - n_{14}}{\Delta}; \quad \vartheta_{12} = \frac{m_{14}n_{44} - m_{44}n_{14}}{\Delta};$$

$$\vartheta_{21} = \frac{n_{13} - n_{43}}{\Delta}; \quad \vartheta_{22} = \frac{m_{44}n_{13} - m_{14}n_{43}}{\Delta}.$$

Expressions (4.72)–(4.74) for $x_1 \to b + 0$ can be simplified to

$$\sigma_{33}^{(1)}(x_1, 0) = \sigma \sqrt{\frac{h}{2\pi (x_1 - b)} \tan \frac{\pi b}{h}},$$

$$\sigma_{13}^{(1)}(x_1, 0) = \tau \sqrt{\frac{h}{2\pi (x_1 - b)} \tan \frac{\pi b}{h}}, \quad (4.81)$$

$$D_3^{(1)}(x_1, 0) = (d - D) \sqrt{\frac{h}{2\pi (x_1 - b)} \tan \frac{\pi b}{h}}.$$

Introducing the energy release rate at the point $x_1 = b$ according to [3] as

$$G = \lim_{\Delta l \to 0} \frac{1}{2\Delta l} \int_b^{b+\Delta l} \left\{ \begin{array}{l} \sigma_{33}^{(1)}(x_1, 0)\langle u_3(x_1 - \Delta l, 0) \rangle + \\ + \sigma_{13}^{(1)}(x_1, 0)\langle u_1(x_1 - \Delta l, 0) \rangle + D_3^{(1)}(x_1, 0)\langle \varphi(x_1 - \Delta l, 0) \rangle \end{array} \right\} dx_1$$

and inserting Eqs. (4.78)–(4.81), one gets after simple calculations

$$G = \frac{h}{4} \tan \frac{\pi b}{h} \left[\vartheta_{11}\sigma^2 + (\vartheta_{12} + \vartheta_{21})\sigma(d - D) + \vartheta_{22}(d - D)^2 - \frac{m_{11}}{n_{11}}\tau^2 \right].$$

Analysis of the results. Consider the influence of the mechanical and electrical loading on the intensity factors and the energy release rate for various electric permittivity of the crack filler. The piezoelectric material PZT-4 is chosen for calculations. The crack length is considered to be tenth part of a period which equals π.

Tables 4.3 and 4.4 show the results for the electric displacement D inside the cracks, the electric intensity factor K_4 and the energy release rate G for the case of a strong mechanical loading and a weak electric displacement applied in the positive and the negative direction, respectively.

The electromechanical state is analyzed in the vicinity of the crack tip in case of a strong electrical loading and a weak mechanical loading (Table 4.5). Particularly strong influence is observed for the electric permittivity values ε_r smaller than 1 (corresponding to the case when the cracks are filled with a substance with low electric permittivity).

Tables 4.6 and 4.7 show the results for the mechanical and electrical loading, which is 10 times stronger than in Tables 4.3 and 4.4, respectively. Since the problem is nonlinear, the proportional change of the intensity factor K_4 and the energy release rate G is only valid for limiting cases of the electrically permeable ($\varepsilon_r = 4000$) and electrically insulating ($\varepsilon_r = 10^{-6}$) cracks.

The variation of the energy release rate G with the ratio of the crack length to the period at the different values of electric permittivity of cracks is presented

Table 4.3 Variation of the electric displacement, the electric intensity factor and the energy release rate under strong mechanical ($\sigma = 10$ MPa) and weak electrical ($d = 0.001$ C/m^2) loading

ε_r	D (C/m^2)	$K_4 \times 10^4$ (C/m$^{3/2}$)	G (N/m)
10^{-6}	-6×10^{-9}	7.05	524.191
1	-0.00119	15.4	572.478
2.5	-0.00137	16.7	574.462
81	-0.00152	17.8	575.022
4000	-0.00153	17.8	575.023

Table 4.4 Variation of the electric displacement, the electric intensity factor and the energy release rate under strong mechanical ($\sigma = 10$ MPa) and weak electrical ($d = -0.001$ C/m^2) loading

ε_r	D (C/m^2)	$K_4 \times 10^4$ (C/m$^{3/2}$)	G (N/m)
10^{-6}	-1.7879×10^{-8}	-7.0538	303.787
1	-0.00276356	12.4401	562.321
2.5	-0.00316293	15.2572	572.136
81	-0.00351383	17.7324	575.019
4000	-0.00352648	17.8216	575.023

Table 4.5 Variation of the electric displacement, the electric intensity factor and the energy release rate under weak mechanical ($\sigma = 1$ MPa) and strong electrical ($d = 0.03$ C/m^2) loading

ε_r	D (C/m^2)	$K_4 \times 10^4$ (C/m$^{3/2}$)	G (N/m)
10^{-6}	3.38×10^{-8}	211.6	-19291.7
1	0.0265	24.67	-223.879
2.5	0.0292	5.47	-0.2095
81	0.0297	1.86	5.7476
4000	0.0297	1.78	5.7502

Table 4.6 Variation of the electric displacement, the electric intensity factor and the energy release rate under the loading $\sigma = 100$ MPa, $d = 0.01$ C/m^2 (10 times stronger than used in Table 4.3)

ε_r	D (C/m^2)	$K_4 \times 10^4$ (C/m$^{3/2}$)	$G \times 10^{-2}$ (N/m)
10^{-6}	-6×10^{-9}	70.5	524.191
1	-0.00417	99.9	548.159
2.5	-0.00727	121.8	561.070
81	-0.0147	174.4	574.959
4000	-0.0153	178.2	575.023

Table 4.7 Variation of the electric displacement, the electric intensity factor and the energy release rate under the loading $\sigma = 100$ MPa, $d = -0.01$ C/m^2 (10 times stronger than used in Table 4.4)

ε_r	D (C/m^2)	$K_4 \times 10^4$ (C/m$^{3/2}$)	$G \times 10^{-2}$ (N/m)
10^{-6}	-1.7879×10^{-8}	-70.5392	303.785
1	-0.0107537	5.31622	443.977
2.5	-0.0177109	54.3921	507.805
81	-0.0340305	169.509	574.689
4000	-0.0352412	178.049	575.022

in Fig. 4.18. A rapid increase in the energy release rate with decreasing distance between the cracks can be observed. It can be seen as well that the value of G is larger for the crack filler with higher electric permittivity.

The influence of the external electromechanical loading on the energy release rate G is presented in Fig. 4.19. It is easy to see that under proportional increase of the electromechanical loading, G varies disproportionately for cracks with finite permittivity. The sections of the proportional variation of G are close to the extreme values of the electric permittivity coefficient. For example, in case of 10 times increase of the external electromechanical loading for the cracks filled with air ($\varepsilon_r = 1$) the energy release rate change is smaller than 10^2 times that would occur in the limiting cases of the electrically permeable or electrically insulating cracks (Fig. 4.19).

The results for the energy release rate G under the constant mechanical and variable electrical loading for the different values of the coefficient of relative elec-

Fig. 4.18 Dependence of
the energy release rate on the
ratio of the crack length to
the period for the different
values of electric permittivity
of cracks

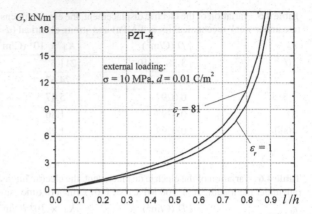

Fig. 4.19 Variation of the
energy release rate under
proportional change of the
electromechanical loading
for the different values of
electric permittivity of the
cracks

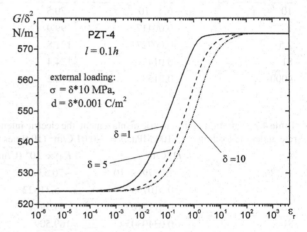

tric permittivity are shown in Fig. 4.20. The most drastic change of the ERR is observed for the cracks with low electric permittivity, while change of the ERR is less significant for more electrically permeable cracks. The energy release rate is negative for the electrically insulating crack due to the strong singularity. In case of the finite electric permittivity, e.g. when the cracks are filled with the air ($\varepsilon_r = 1$), the singularity of the electric field is much smaller and the ERR takes positive values.

It should be noted that the approximate approach which is based on the assumption of the electrical isolation of the crack leads to a contradiction when using the ERR as a fracture criterion: according to it, fracture will not occur because there is more accumulated energy than released. On the contrary, when taking into account the finite permittivity, the energy is released as crack grows. It should be noted that the ERR decreases with increasing the electrical loading regardless of the sign. This indicates the restraining contribution of the electrical loading on the fracture of piezoelectric materials.

Conclusions. The problem for a periodic set of cracks of finite electric permittivity located in a homogeneous piezoelectric space is considered. Analytical formulae

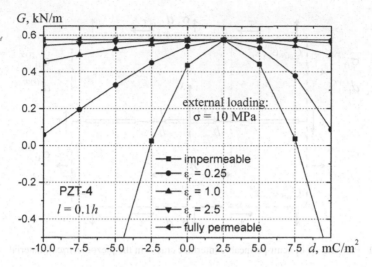

Fig. 4.20 Dependence of the energy release rate on the external electrical loading for the different values of electric permittivity of the cracks

are obtained for the stresses and the electrical displacement distribution. The stress intensity factors K_1, K_2 and the electric intensity factor K_4 are derived. The energy release rate G is computed as well. Numerical calculations are carried out for the piezoelectric PZT-4. It is found that the results under the low electrical loading are similar to those obtained in the case of electrically permeable cracks. It is found that the dependence of the electric displacement inside the cracks, the intensity factors and the energy release rate nonlinearly depend on the applied electromechanical loading.

4.4.2 Solution to the Problem Within the Classical Model

A solution to the problem of a periodic set of cracks with finite electrical permeability located at the interface of two dissimilar piezoelectric materials is constructed within the classical model in this subsection.

Statement of the problem. Consider a periodic set of cracks located at the interface of two dissimilar piezoelectric half-spaces [10]. The characteristics of the components are assigned by their elastic moduli $C_{ijkl}^{(m)}$, piezoelectric constants $e_{ikl}^{(m)}$ and dielectric constants $\varepsilon_{il}^{(m)}$. Superscript "1" ("2") corresponds to the "upper" ("lower") material.

Uniformly distributed tensile (σ) and shear (τ) stresses are assumed to act at infinity and a uniformly distributed electric displacement d is assigned as well. The cracks are opened under the action of these factors. The coordinates of the crack tips in a fundamental strip [2] are denoted by $-b$ and b, and a period is assumed to be

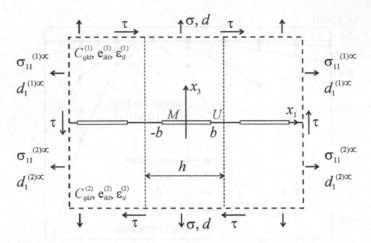

Fig. 4.21 Periodic set of semi-permeable interface cracks in the piezoelectric bimaterial

equal to h (Fig. 4.21). In addition, the union of the cracks is denoted by M, and the union of the bonded parts of the interface is denoted by U.

Similarly to the case of a homogeneous material, described in the preceding subsection, condition (4.53) is obtained inside the cracks. Assuming the open parts of the cracks to be load-free, the continuity and boundary conditions can be written in the form

$$\langle \mathbf{V}(x_1, 0)\rangle = 0, \quad \langle \mathbf{t}(x_1, 0)\rangle = 0, \quad x_1 \in U, \tag{4.82}$$

$$\sigma_{13}^{\pm}(x_1, 0) = 0, \quad \langle \sigma_{33}(x_1, 0)\rangle = 0,$$
$$D_3\langle u_3(x_1, 0)\rangle = -\varepsilon_a\langle\varphi(x_1, 0)\rangle, \quad \langle D_3(x_1, 0)\rangle = 0, \quad x_1 \in M. \tag{4.83}$$

General analytical solution. For the considered loading and the bimaterial components made of 6 mm class ceramics polarized along the x_3 direction, a plane strain state takes place. After transformations similar to Sect. 4.4.1, the following expressions for the stresses and the electric displacement at the interface can be derived:

$$\sigma_{33}^{(1)}(x_1, 0) + m_{j4}D_3^{(1)}(x_1, 0) + im_{j1}\sigma_{13}^{(1)}(x_1, 0) = F_j^+(x_1) + \gamma_j F_j^-(x_1), \tag{4.84}$$

where

$$F_j(z) = n_{j1}W_1(z) + i[n_{j3}W_3(z) + n_{j4}W_4(z)], \tag{4.85}$$

and the coefficients m_{jl}, n_{jl}, γ_j $(j, l = 1, 3, 4)$ depend on the electromechanical properties of the composite and are real for the piezoceramic of class 6 mm.

The electric displacement is assumed to be a constant inside the cracks, namely

$$D_3^+(x_1, 0) = D_3^-(x_1, 0) = D, \quad x_1 \in M. \tag{4.86}$$

Using Eqs. (4.84) and (4.86) together with conditions (4.83) at the crack faces, the following periodic Riemann problem is obtained:

$$F_j^+(x_1) + \gamma_j F_j^-(x_1) = m_{j4}D, \quad (j = 1, 3, 4), \quad x_1 \in M. \tag{4.87}$$

To get the *homogeneous* periodic Riemann problem, introduce a new function

$$\Phi_j(z) = F_j(z) - m_{j4}D/(1 + \gamma_j). \tag{4.88}$$

Then the periodic Riemann problem (4.87) takes the form

$$\Phi_j^+(x_1) + \gamma_j \Phi_j^-(x_1) = 0, \quad (j = 1, 3, 4), \quad x_1 \in M, \tag{4.89}$$

$$\Phi_j(z)\Big|_{z \to \pm i\infty} = \sigma_j^* - i\tau_j^*, \tag{4.90}$$

where $\sigma_j^* = [\sigma + m_{j4}(d - D)]/(1 + \gamma_j)$, $\tau_j^* = -m_{j1}\tau/(1 + \gamma_j)$.

Based on [2], the solution to the homogeneous periodic Riemann problem (4.89) will be searched in the form

$$\Phi_j(z) = \frac{C_{1j}\cos\frac{\pi z}{h} + C_{2j}\sin\frac{\pi z}{h}}{\sqrt{\sin\frac{\pi(z-b)}{h}\sin\frac{\pi(z+b)}{h}}} \left[\frac{\sin\frac{\pi(z+b)}{h}}{\sin\frac{\pi(z-b)}{h}}\right]^{i\varepsilon_j}. \tag{4.91}$$

Here $\varepsilon_j = \ln\gamma_j/(2\pi)$. After calculations similar to Sect. 4.3.1, the behavior of the function $\Phi_j(z)$ at infinity is determined as

$$\Phi_j(z)\Big|_{z \to +i\infty} = (C_{2j} - iC_{1j})e^{2\pi b\varepsilon_j/h},$$

$$\Phi_j(z)\Big|_{z \to -i\infty} = (C_{2j} + iC_{1j})e^{-2\pi b\varepsilon_j/h}. \tag{4.92}$$

Using formulae (4.90) and (4.92), a system of linear algebraic equations for arbitrary complex constants C_{1j} and C_{2j} is obtained. Solving this system gives

$$C_{1j} = -\sin(2\pi i\varepsilon_j b/h)(\sigma_j^* - i\tau_j^*),$$

$$C_{2j} = \cos(2\pi i\varepsilon_j b/h)(\sigma_j^* - i\tau_j^*).$$

Substituting the obtained values for constants C_{1j} and C_{2j} into Eq. (4.91) and conducting elementary transformations, the solution to the homogeneous periodic Riemann problem (4.89) takes the form

$$\Phi_j(z) = (\sigma_j^* - i\tau_j^*)\frac{\Xi(z - 2i\varepsilon_j b)}{\sqrt{\Xi(z-b)\Xi(z+b)}}\left[\frac{\Xi(z+b)}{\Xi(z-b)}\right]^{i\varepsilon_j}. \tag{4.93}$$

Using Eq. (4.85), the following expression is derived:

$$n_{j1}\langle u_1'(x_1,0)\rangle + i\left(n_{j3}\langle u_3'(x_1,0)\rangle + n_{j4}\langle \varphi'(x_1,0)\rangle\right) = F_j^+(x_1) - F_j^-(x_1).$$

Since $\Phi_j^-(x_1) = -\Phi_j^+(x_1)/\gamma_j$, $x_1 \in M$, then $F_j^+(x_1) - F_j^-(x_1) = (\gamma_j + 1)/\gamma_j \Phi_j^+(x_1)$ and the formula for the derivative of the displacement jump takes the form

$$n_{j1}\langle u_1'(x_1,0)\rangle + i\left(n_{j3}\langle u_3'(x_1,0)\rangle + n_{j4}\langle \varphi'(x_1,0)\rangle\right) =$$

$$= -i\frac{\gamma_j + 1}{\sqrt{\gamma_j}}\left(\sigma_j^* - i\tau_j^*\right)\frac{\varXi(x_1 - 2i\varepsilon_j b)}{\sqrt{\varXi(b-x_1)\varXi(b+x_1)}}\left[\frac{\varXi(b+x_1)}{\varXi(b-x_1)}\right]^{i\varepsilon_j}, \quad x_1 \in M.$$

$$(4.94)$$

After transformations of the last expression, taking into account that ε_1 is small and $\varepsilon_4 = 0$ and using relation (4.53), the following equation is obtained to determine the electric displacement D inside the cracks:

$$D = \varepsilon_a \frac{\gamma_0 n_{43}\sigma_1^*(D) - 2n_{13}\sigma_4^*(D)}{\gamma_0 n_{44}\sigma_1^*(D) - 2n_{14}\sigma_4^*(D)},$$

$$(4.95)$$

where $\gamma_0 = \dfrac{\gamma_1 + 1}{\sqrt{\gamma_1}}$.

Equation (4.95) can be rewritten in the form

$$\eta_1^* D^2 + \eta_2^* D + \eta_3^* = 0,$$

$$(4.96)$$

where

$$\eta_1^* = \sqrt{\gamma_1}m_{44}n_{14} - m_{14}n_{44},$$
$$\eta_2^* = (s_1 n_{44} + \varepsilon_a m_{14}n_{43}) - \sqrt{\gamma_1}(s_4 n_{14} + \varepsilon_a m_{44}n_{13}),$$
$$\eta_3^* = \varepsilon_a(\sqrt{\gamma_1}s_4 n_{13} - s_1 n_{43}).$$

Using Eqs. (4.84), (4.88) and (4.93) for $j = 1$, and the real part of these equations for $j = 4$, the system of equations is obtained to determine the stresses and electric displacement in $x_1 \in U$:

$$\sigma_{33}^{(1)}(x_1,0) + m_{14}D_3^{(1)}(x_1,0) + im_{11}\sigma_{13}^{(1)}(x_1,0) =$$

$$= (\sigma + m_{14}(d - D) + im_{11}\tau)\frac{\varXi(x_1 - 2i\varepsilon_1 b)}{\sqrt{\varXi(x_1 - b)\varXi(x_1 + b)}}\left[\frac{\varXi(x_1 + b)}{\varXi(x_1 - b)}\right]^{i\varepsilon_1} + m_{14}D,$$

$$(4.97)$$

$$\sigma_{33}^{(1)}(x_1,0) + m_{44}D_3^{(1)}(x_1,0) = (\sigma + m_{44}(d - D))\frac{\varXi(x_1)}{\sqrt{\varXi(x_1 - b)\varXi(x_1 + b)}} + m_{44}D.$$

$$(4.98)$$

Introduce the intensity factor (IF) similar to [11]:

$$K_1^{osc} + m_{14}K_4^{osc} + im_{11}K_2^{osc} =$$
$$= \lim_{x_1 \to b+0} \sqrt{2\pi \, \Xi(x_1 - b)} \, \Xi(x_1 - b)^{i\varepsilon_1} \left[\sigma_{33}^{(1)}(x_1, 0) + m_{14}D_3^{(1)}(x_1, 0) + im_{11}\sigma_{13}^{(1)}(x_1, 0) \right],$$
(4.99)

$$K_1^{osc} + m_{44}K_4^{osc} = \lim_{x_1 \to b+0} \sqrt{2\pi \, \Xi(x_1 - b)} \left[\sigma_{33}^{(1)}(x_1, 0) + m_{44}D_3^{(1)}(x_1, 0) \right].$$
(4.100)

Using Eqs. (4.97), (4.98) as $x_1 \to b + 0$, the following system of equations is obtained to determine the IFs:

$$K_1^{osc} + m_{14}K_4^{osc} + im_{11}K_2^{osc} = \sqrt{2\pi}(\sigma + m_{14}(d - D) + im_{11}\tau) \frac{\Xi(b - 2i\varepsilon_1 b)}{\sqrt{\Xi(2b)}} [\Xi(2b)]^{i\varepsilon_1},$$

$$K_1^{osc} + m_{44}K_4^{osc} = \sqrt{\pi \tan \frac{\pi b}{h}}(\sigma + m_{44}(d - D)).$$

Solving this system one gets:

$$K_1^{osc} = \sqrt{2\pi} \frac{m_{44}\{\Omega_1 [\sigma + m_{14}(d - D)] + \Omega_2 m_{11}\tau\} - m_{14}\Xi(b)(\sigma + m_{44}(d - D))}{(m_{44} - m_{14})\sqrt{\Xi(2b)}},$$
(4.101)

$$K_2^{osc} = \frac{\sqrt{2\pi}}{m_{11}} \sqrt{\frac{1}{\Xi(2b)}}\{\Omega_1 m_{11}\tau - \Omega_2 [\sigma + m_{14}(d - D)]\},$$
(4.102)

$$K_4^{osc} = \sqrt{2\pi} \frac{\Xi(b)(\sigma + m_{44}(d - D)) - \{\Omega_1 [\sigma + m_{14}(d - D)] + \Omega_2 m_{11}\tau\}}{(m_{44} - m_{14})\sqrt{\Xi(2b)}}.$$
(4.103)

where
$$\Omega_1 = \sin \frac{\pi b}{h} \times \cosh \frac{2\pi \varepsilon_1 b}{h} \cos \zeta + \cos \frac{\pi b}{h} \times \sinh \frac{2\pi \varepsilon_1 b}{h} \sin \zeta,$$
$$\Omega_2 = \cos \frac{\pi b}{h} \times \sinh \frac{2\pi \varepsilon_1 b}{h} \cos \zeta - \sin \frac{\pi b}{h} \times \cosh \frac{2\pi \varepsilon_1 b}{h} \sin \zeta, \quad \zeta = \varepsilon_1 \ln \sin 2b.$$

Analysis of the results. Consider a periodic set of cracks located at the interface of two piezoelectric materials PZT-4 and PZT-5. Let us study the influence of the mechanical and electrical loading onto the electric displacement inside the cracks and onto the intensity factors for various values of the electric permittivity coefficient. The crack length is tenth of a period which equals π. It should be noted that the electric permittivity coefficient ε_r is equal to 1 for air, 2.5 for silicone oil and 81 for water. The values of ε_r, which are equal to 10^{-6} and 4000, can be considered corresponding to the electrically insulating and fully electrically permeable cracks, respectively.

Table 4.8 Variation of the electric displacement and the electric IF under strong mechanical ($\sigma =$ 10 MPa) and weak electrical ($d = 0.001$ C/m^2) loading

ε_r	D (C/m^2)	K_1^{osc} (MPa/$\sqrt{m}$)	K_2^{osc} (MPa/$\sqrt{m}$)	$K_4^{osc} \times 10^4$ (C/m$^{3/2}$)
10^{-6}	-5.72×10^{-9}	7.05319	0.131331	7.05494
1	-0.00130	7.05303	0.159217	16.2392
2.5	-0.00154	7.05300	0.164234	17.8916
81	-0.00175	7.05298	0.168734	19.3734
4000	-0.00175	7.05298	0.168897	19.4272

Table 4.9 Variation of the electric displacement and the electric IF under strong mechanical ($\sigma =$ 10 MPa) and weak electrical ($d = -0.001$ C/m^2) loading

ε_r	D (C/m^2)	K_1^{osc} (MPa/$\sqrt{m}$)	K_2^{osc} (MPa/$\sqrt{m}$)	$K_4^{osc} \times 10^4$ (C/m$^{3/2}$)
10^{-6}	-1.53552×10^{-8}	7.05343	0.0884938	-7.05317
1	-0.00281801	7.05309	0.148851	12.8252
2.5	-0.00329645	7.05303	0.159099	16.2002
81	-0.00373743	7.05298	0.168544	19.3109
4000	-0.00375374	7.05298	0.168893	19.4259

The results of the numerical calculations are presented in Tables 4.8 and 4.9 for the cases of a strong mechanical loading and a weak electric displacement acting in the positive and negative directions, respectively.

In both cases, the mechanical SIF K_1^{osc} varies only slightly for the different values of the electric permittivity coefficient, while K_2^{osc} changes moderately. The electrical intensity factor K_4^{osc} depends significantly on the electric permittivity of the cracks. For the cracks with high electric permittivity the mechanical loading makes the greatest contribution to the K_4^{osc} value. It can be also seen from the fact that K_4^{osc} remain almost the same when the direction of the applied electrical loading changes to the opposite one. The influence of the remote electric displacement becomes essentially greater with decreasing the electric permittivity of the cracks. For the above loading case, the values of K_4^{osc} are almost equal on the absolute value but have the opposite sign for the electrically insulating crack.

The dependence of the electrical IF K_4^{osc} on the electrical loading is shown in Fig. 4.22. Here, the electrical loading continuously varies from -0.01 to 0.01 C/m^2 for the different values of the relative electric permittivity coefficient ε_r.

K_4^{osc} remains almost constant for the fully electrically permeable cracks. The influence of the external electrical loading is more significant with decreasing ε_r because lesser electrical conductivity leads to the greater electrical singularity in the vicinity of the crack tips due to more pronounced shielding. K_4^{osc} is equal to zero under zero external electrical loading only for the case of electrically insulating

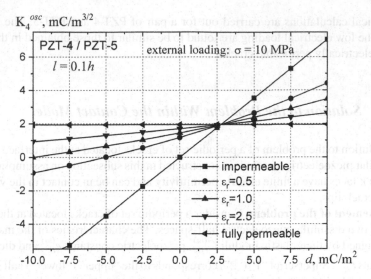

Fig. 4.22 Dependence of the electric IF K_4^{osc} on the external electrical loading plotted for the different values of electric permittivity of the cracks

crack. In other cases, the mechanical loading affects the electric singularity as the component of the electric displacement, normal to the crack faces, is non-zero. All curves are concurrent at $d = 0.0025$ C/m^2, i.e. the K_4^{osc} value under such loading coincide for cracks of any electric permittivity.

The results under a weak mechanical loading and a strong electric displacement are shown in Table 4.10. It should be noted that when ε_r is approaching zero (electrically insulating crack), the IFs change abruptly. It is due to the singularity of the stress fields in the vicinity of the crack tips and the strong electrical loading.

Conclusions. The problem for a periodic set of the cracks of finite electric permittivity located at the interface of two piezoelectric materials is considered. Analytical formulae are obtained for the stresses and the electric displacement, as well as for the mechanical stress intensity factors K_1^{osc}, K_2^{osc} and the electric intensity factor K_4^{osc}.

Table 4.10 Variation of the electric displacement and the electric IF under weak mechanical ($\sigma = 1000$ Pa) and strong electrical ($d = 0.01$ C/m^2) loading

ε_r	D (C/m^2)	K_1^{osc} (MPa/$\sqrt{m}$)	K_2^{osc} (MPa/$\sqrt{m}$)	$K_4^{osc} \times 10^4$ (C/m$^{3/2}$)
10^{-6}	3.21×10^{-8}	-0.5039	214.2	70.5407
1	0.00999	0.7052	0.0281	0.00564
2.5	0.00999	0.7053	0.0204	0.00311
81	0.00999	0.7053	0.0170	0.00197
4000	0.00999	0.7053	0.0169	0.00194

Numerical calculations are carried out for a pair of PZT-4 and PZT-5. The results under the low electrical loading are found to be similar to those obtained in the case of the electrically permeable cracks.

4.4.3 Solution to the Problem Within the Contact Model

The solution to the problem of a periodic set of cracks located at the interface of two dissimilar piezoelectric materials is constructed in this subsection on assumption that the crack faces have a finite electric permittivity and can be in contact in the vicinity of the crack tips.

Statement of the problem. Consider a periodic set of cracks located at the interface of two dissimilar piezoelectric half-spaces. The characteristics of the materials are assigned by their elastic moduli $C_{ijkl}^{(m)}$, piezoelectric constants $e_{ikl}^{(m)}$ and dielectric constants $\varepsilon_{il}^{(m)}$. Superscript "1" ("2") corresponds to the "upper" ("lower") half-space.

Uniformly distributed tensile (σ) and shear (τ) stresses are assumed to act at infinity and a uniformly distributed electric displacement d is assigned as well (Fig. 4.23). The cracks are opened under the action of these factors. The tips of the open crack part in the fundamental strip [2] are denoted by c and b. The point between the contact zone and the bonded part of the interface is denoted by a. The period is assumed to be equal h. In addition, the union of the open crack faces is denoted by M, the union of the contact zones - by L and the union of the bonded parts of the interface - by U.

As in the preceding subsection, condition (4.53) is used at the open parts of the cracks. Assuming that the open crack parts are load-free and the contact zones are smooth and electrically permeable, the continuity and boundary conditions can be written in the form

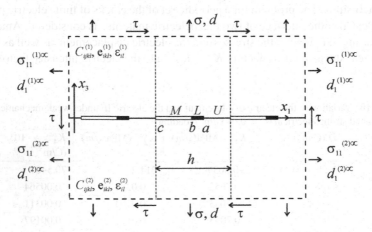

Fig. 4.23 Periodic set of semi-permeable interface cracks with contact zones in the piezoelectric bimaterial

$$\langle \mathbf{V}(x_1, 0) \rangle = 0, \quad \langle \mathbf{t}(x_1, 0) \rangle = 0, \quad x_1 \in U, \tag{4.104}$$

$$\sigma_{13}^{(m)}(x_1, 0) = 0, \quad \langle \sigma_{33}(x_1, 0) \rangle = 0, \quad \langle D_3(x_1, 0) \rangle = 0,$$
$$\langle u_3(x_1, 0) \rangle = 0, \quad \langle \varphi(x_1, 0) \rangle = 0, \quad x_1 \in L, \tag{4.105}$$

$$\sigma_{13}^{(m)}(x_1, 0) = 0, \quad \sigma_{33}^{(m)}(x_1, 0) = 0, \quad \langle D_3(x_1, 0) \rangle = 0,$$
$$D_3(x_1, 0)\langle u_3(x_1, 0) \rangle = -\varepsilon_a \langle \varphi(x_1, 0) \rangle, \quad x_1 \in M. \tag{4.106}$$

General analytical solution. The components of the bimaterial are assumed to be piezoelectrics of class 6mm polarized in the direction of the x_3-axis. A plane strain takes place for the considered loading.

The electric induction is assumed to be constant along the crack faces:

$$D_3^+(x_1, 0) = D_3^-(x_1, 0) = D, \quad x_1 \in M. \tag{4.107}$$

Using Eqs. (4.84) and (4.85) together with the conditions at the open and bonded parts of the interface, the following equations are obtained:

$$F_j^+(x_1) + \gamma_j F_j^-(x_1) = m_{j4}D, \quad x_1 \in M,$$
$$\mathrm{Im}\left(F_j^+(x_1) - F_j^-(x_1)\right) = 0, \quad x_1 \in L, \quad (j = 1, 3, 4) \tag{4.108}$$
$$\mathrm{Im}\left(F_j^+(x_1) + \gamma_j F_j^-(x_1)\right) = 0, \quad x_1 \in L,$$

or, gathering Eqs. (4.108$_2$) and (4.108$_3$), the following combined periodic Dirichlet-Riemann problem is obtained:

$$\begin{cases} F_j^+(x_1) + \gamma_j F_j^-(x_1) = m_{j4}D, & x_1 \in M, \\ \mathrm{Im}\, F_j^{\pm}(x_1) = 0, & x_1 \in L. \end{cases} \quad (j = 1, 3, 4) \tag{4.109}$$

Behaviour of the function $F_j(z)$ at infinity is the following:

$$F_j(z)\Big|_{z \to \pm i\infty} = \tilde{\sigma}_j - i\tilde{\tau}_j,$$

$$\tilde{\sigma}_j = \frac{\sigma + m_{j4}d}{1 + \gamma_j}, \quad \tilde{\tau}_j = -\frac{m_{j1}\tau}{1 + \gamma_j}.$$

To obtain the homogeneous Riemann problem, let us introduce a new function

$$\Phi_j(z) = F_j(z) - \frac{m_{j4}D}{1 + \gamma_j}. \tag{4.110}$$

As a result, Eq. (4.109) is simplified to the combined homogeneous Dirichlet-Riemann problem

$$\begin{cases} \Phi_j^+(x_1) + \gamma_j \Phi_j^-(x_1) = 0, & x_1 \in M, \\ \mathrm{Im}\, \Phi_j^\pm(x_1) = 0, & x_1 \in L, \end{cases} \quad (j = 1, 3, 4), \tag{4.111}$$

$$\Phi_j(z)\Big|_{z \to \pm i\infty} = \sigma_j^* - i\tau_j^*, \tag{4.112}$$

where

$$\sigma_j^* = \frac{\sigma + m_{j4}(d - D)}{1 + \gamma_j}, \quad \tau_j^* = \tilde{\tau}_j = -\frac{m_{j1}\tau}{1 + \gamma_j}.$$

General solution to this problem is constructed in Chap. 2. It has the following form:

$$\Phi_j(z) = \frac{e^{i\varphi_j(z)}}{\sqrt{\Xi(z - c)}} \left(\frac{P_j(z)}{\sqrt{\Xi(z - b)}} + i \frac{Q_j(z)}{\sqrt{\Xi(z - a)}} \right), \tag{4.113}$$

where

$$\varphi_j(z) = 2\varepsilon_j \ln \left(\frac{\sqrt{\Xi(a - b)\Xi(z - c)}}{\sqrt{\Xi(a - c)\Xi(z - b)} + \sqrt{\Xi(b - c)\Xi(z - a)}} \right),$$

$$P_j(z) = C_{1j} \cos[\pi(z - a_*)/h] + C_{2j} \sin[\pi(z - a_*)/h], \quad a_* = (c + b)/2,$$

$$Q_j(z) = D_{1j} \cos[\pi(z - b_*)/h] + D_{2j} \sin[\pi(z - b_*)/h], \quad b_* = (c + a)/2,$$

$$\varepsilon_j = \ln \gamma_j / (2\pi),$$

$C_{1j}, C_{2j}, D_{1j}, D_{2j}$ are arbitrary real constants, which are determined by the behavior of the function at infinity (as $z \to \pm i\infty$).

Considering the behavior of function (4.113) at infinity (cf. Chap. 2) and using Eq. (4.112), a system of linear algebraic equations is obtained:

$$\begin{cases} e^{\chi_j}(\cos \zeta_j - i \sin \zeta_j)\left[C_{2j} - iC_{1j} + D_{1j} + iD_{2j} \right] = \sigma_j^* - i\tau_j^*, \\ e^{-\chi_j}(\cos \zeta_j - i \sin \zeta_j)\left[C_{2j} + iC_{1j} - D_{1j} + iD_{2j} \right] = \sigma_j^* - i\tau_j^*, \end{cases} \tag{4.114}$$

in which

$$\chi_j = 2\varepsilon_j \arctan \frac{\sqrt{\sin \frac{\pi(a-c)}{h}} \sin \frac{\pi(b-c)}{2h} + \sqrt{\sin \frac{\pi(b-c)}{2h}} \sin \frac{\pi(a-c)}{2h}}{\sqrt{\sin \frac{\pi(a-c)}{h}} \cos \frac{\pi(b-c)}{2h} + \sqrt{\sin \frac{\pi(b-c)}{h}} \cos \frac{\pi(a-c)}{2h}},$$

$$\zeta_j = \varepsilon_j \ln \frac{\sin \frac{\pi(a+b-2c)}{2h} + \sqrt{\sin \frac{\pi(a-c)}{h} \sin \frac{\pi(b-c)}{h}}}{\sin \frac{\pi(a-b)}{2h}}.$$

Solving system (4.114), the real constants are determined as

$$C_{1j} = \sinh \chi_j \left(-\tau_j^* \cos \zeta_j + \sigma_j^* \sin \zeta_j\right), \quad C_{2j} = \cosh \chi_j \left(\sigma_j^* \cos \zeta_j + \tau_j^* \sin \zeta_j\right),$$

$$D_{1j} = -\sinh \chi_j \left(\sigma_j^* \cos \zeta_j + \tau_j^* \sin \zeta_j\right), \quad D_{2j} = \cosh \chi_j \left(-\tau_j^* \cos \zeta_j + \sigma_j^* \sin \zeta_j\right).$$

Using Eq. (4.85), the following expression is derived:

$$n_{j1}\langle u_1'(x_1, 0)\rangle + i \left(n_{j3}\langle u_3'(x_1, 0)\rangle + n_{j4}\langle \varphi'(x_1, 0)\rangle\right) = F_j^+(x_1) - F_j^-(x_1).$$

Since $\Phi_j^-(x_1) = -\Phi_j^+(x_1)/\gamma_j$, $x_1 \in M$, then $F_j^+(x_1) - F_j^-(x_1) = (\gamma_j + 1)/\gamma_j \Phi_j^+(x_1)$ and the formula for the derivative of the displacement jump takes the form

$$n_{j1}\langle u_1'(x_1, 0)\rangle + i \left(n_{j3}\langle u_3'(x_1, 0)\rangle + n_{j4}\langle \varphi'(x_1, 0)\rangle\right) = \frac{\gamma_j + 1}{\gamma_j}\Phi_j^+(x_1), \quad x_1 \in M.$$

Using formula (4.113) for the open part of the crack, it follows that

$$n_{j3}\langle u_3'(x_1, 0)\rangle + n_{j4}\langle \varphi'(x_1, 0)\rangle = \tag{4.115}$$

$$= \frac{-2\cosh(\pi\varepsilon_j)}{\sqrt{\Xi(x_1 - c)}} \left(\frac{P_j(x_1)}{\sqrt{\Xi(b - x_1)}} \cos \varphi_j^*(x_1) - \frac{Q_j(x_1)}{\sqrt{\Xi(a - x_1)}} \sin \varphi_j^*(x_1)\right), \quad x_1 \in M,$$

$$\varphi_j^*(x_1) = 2\varepsilon_j \ln \frac{\sqrt{\Xi(a - b)\Xi(x_1 - c)}}{\sqrt{\Xi(a - c)\Xi(b - x_1)} + \sqrt{\Xi(b - c)\Xi(a - x_1)}}, \quad x_1 \in M.$$

Considering the previous expression for $j = 1, 4$ and taking into account that $C_{14} = 0$, $C_{24} = \sigma_4^*$, the following system is obtained:

$$\begin{cases} n_{13}\langle u_3'(x_1, 0)\rangle + n_{14}\langle \varphi'(x_1, 0)\rangle = \\ = \dfrac{-2\cosh(\pi\varepsilon_1)}{\sqrt{\Xi(x_1 - c)}} \left(\dfrac{P_1(x_1)}{\sqrt{\Xi(b - x_1)}} \cos \varphi_1^*(x_1) - \dfrac{Q_1(x_1)}{\sqrt{\Xi(a - x_1)}} \sin \varphi_1^*(x_1)\right), \quad x_1 \in M, \\ n_{43}\langle u_3'(x_1, 0)\rangle + n_{44}\langle \varphi'(x_1, 0)\rangle = \dfrac{-2\sigma_4^* \Xi(x_1 - \frac{c+b}{2})}{\sqrt{\Xi(x_1 - c)\Xi(b - x_1)}}, \quad x_1 \in M. \end{cases}$$

The expressions for the jump of the derivative of the normal displacement component and electric potential on $x_1 \in M$ follow from this system:

$$\langle u_3'(x_1, 0)\rangle = \frac{1}{n_{13}n_{44} - n_{43}n_{14}} \left\{ n_{14} \frac{2\sigma_4^* \Xi(x_1 - \frac{c+b}{2})}{\sqrt{\Xi(x_1 - c)\Xi(b - x_1)}} - n_{44} \frac{2\cosh(\pi\varepsilon_1)}{\sqrt{\Xi(x_1 - c)}} \times \right.$$

$$\left. \times \left(\frac{P_1(x_1)}{\sqrt{\Xi(b - x_1)}} \cos \varphi_1^*(x_1) - \frac{Q_1(x_1)}{\sqrt{\Xi(a - x_1)}} \sin \varphi_1^*(x_1)\right) \right\}, \tag{4.116}$$

$$\langle \varphi'(x_1,0) \rangle = \frac{1}{n_{13}n_{44} - n_{43}n_{14}} \left\{ - n_{13} \frac{2\sigma_4^* \mathcal{E}(x_1 - \frac{c+b}{2})}{\sqrt{\mathcal{E}(x_1-c)\mathcal{E}(b-x_1)}} + n_{43} \frac{2\cosh(\pi \varepsilon_1)}{\sqrt{\mathcal{E}(x_1-c)}} \times \right.$$

$$\left. \times \left(\frac{P_1(x_1)}{\sqrt{\mathcal{E}(b-x_1)}} \cos \varphi_1^*(x_1) - \frac{Q_1(x_1)}{\sqrt{\mathcal{E}(a-x_1)}} \sin \varphi_1^*(x_1) \right) \right\}. \qquad (4.117)$$

After transformations of the last two expressions, taking into account that ε_1 is small and $\varepsilon_4 = 0$ and using condition (4.53), Eqs. (4.95), (4.96) are obtained to determine the electrical displacement D inside the cracks.

It was assumed by Eq. (4.107) that the electric displacement D is uniform inside the cracks. This assumption was used earlier for an interface crack in a piezoelectric bimaterial by Govorukha et al. [11], Li and Chen [12]. For a crack in a homogeneous piezoelectric media this assumption is completely valid, whereas for an interface crack there is some nonuniformity in the electric field distribution, especially near the crack tips. It was shown in [11] that for a single interface crack this nonuniformity is rather insensitive. In Fig. 4.24 the correctness of the assumption of the constant D value along the crack is illustrated for the considered here periodic case.

Hereafter, the period is taken π for the numerical analysis. In Fig. 4.24 the comparison of the electric field D_3 along the cracks for the electrically permeable interface cracks and the constant D value obtained from Eq. (4.96) for $\varepsilon_r = 4000$ are presented. Piezoelectric/dielectric and piezoelectric/piezoelectric material combinations PZT-PIC151/SiC and PZT-4/PZT-5 are used for different values of the applied electrical and mechanical loading. The length of the crack is taken as one fourth of the period. More detailed view at the vicinity of the crack tip is illustrated in insert (the left part of Fig. 4.24). It is seen that, even for the most severe cases of electromechanical loading, electric displacement field is almost uniform along the cracks and agree well with the constant value obtained from the solution of the Eq. (4.96).

Using Eqs. (4.84), (4.110) and (4.113), the expression for the stresses and electric displacement ahead of the crack tip has the following form:

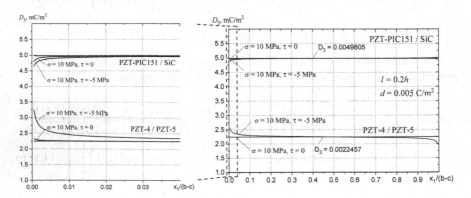

Fig. 4.24 Distribution of the electric displacement field inside the crack (right), including detailed information for the region near the crack tip (left)

$$\sigma_{33}^{(1)}(x_1, 0) + m_{j4}D_3^{(1)}(x_1, 0) + im_{j1}\sigma_{13}^{(1)}(x_1, 0) =$$

$$= (1 + \gamma_j)\frac{e^{i\varphi_j(z)}}{\sqrt{\Xi(z-c)}}\left(\frac{P_j(z)}{\sqrt{\Xi(z-b)}} + i\frac{Q_j(z)}{\sqrt{\Xi(z-a)}}\right) + m_{j4}D, \quad x_1 \in U.$$

$$(4.118)$$

Considering this expression for $j = 1$, as well as its real part for $j = 4$, and taking into account that $C_{14} = 0$, $C_{24} = \sigma_4^*$, $D_{14} = 0$, $D_{24} = -\tau_4^*$, the following system of equations is derived to determine stresses and electric displacement on $x_1 \in U$:

$$\begin{cases} \sigma_{33}^{(1)}(x_1, 0) + m_{14}D_3^{(1)}(x_1, 0) + im_{11}\sigma_{13}^{(1)}(x_1, 0) = \\ = (1 + \gamma_1)\frac{e^{i\varphi_1(x_1)}}{\sqrt{\Xi(x_1-c)}}\left(\frac{P_1(x_1)}{\sqrt{\Xi(x_1-b)}} + i\frac{Q_1(x_1)}{\sqrt{\Xi(x_1-a)}}\right) + m_{14}D, \quad x_1 \in U. \\ \sigma_{33}^{(1)}(x_1, 0) + m_{44}D_3^{(1)}(x_1, 0) = \dfrac{2\sigma_4^*\Xi(x_1 - \frac{c+b}{2})}{\sqrt{\Xi(x_1-c)\Xi(x_1-b)}} + m_{44}D, \end{cases}$$

$$(4.119)$$

The intensity factors at the point a are introduced by the formulae

$$K_{1a} = \lim_{x_1 \to a+0} \sqrt{2\pi\Xi(x_1-a)}\sigma_{33}^{(1)}(x_1, 0),$$

$$K_{2a} = \lim_{x_1 \to a+0} \sqrt{2\pi\Xi(x_1-a)}\sigma_{13}^{(1)}(x_1, 0), \qquad (4.120)$$

$$K_{4a} = \lim_{x_1 \to a+0} \sqrt{2\pi\Xi(x_1-a)}D_3^{(1)}(x_1, 0).$$

Using expressions for the stresses and electric displacement, the intensity factors are found from the following system:

$$\begin{cases} K_{1a} + m_{14}K_{4a} + im_{11}K_{2a} = \sqrt{2\pi}(1+\gamma_1)\dfrac{iQ_1(a)}{\sqrt{\Xi(a-c)}}, \\ K_{1a} + m_{44}K_{4a} = 0. \end{cases}$$

Finally, one gets

$$K_{1a} = 0, \quad K_{4a} = 0, \quad K_{2a} = \frac{(1+\gamma_1)Q_1(a)}{m_{11}\sqrt{\Xi(a-c)}}. \qquad (4.121)$$

The stresses and electric displacement in the contact zone are defined by

$$\sigma_{33}^{(1)}(x_1, 0) + m_{j4}D_3^{(1)}(x_1, 0) + im_{j1}\sigma_{13}^{(1)}(x_1, 0) = \Phi_j^+(x_1) + \gamma_j\Phi_j^-(x_1) + m_{j4}D, \quad x_1 \in L$$

or, using Eq. (4.113) in the contact zone region,

$$\sigma_{33}^{(1)}(x_1, 0) + m_{j4}D_3^{(1)}(x_1, 0) = m_{j4}D + \frac{2e^{\pi\varepsilon_j}}{\sqrt{\Xi(x_1 - c)}} \times \tag{4.122}$$

$$\times \left(\frac{P_j(x_1)}{\sqrt{\Xi(x_1 - b)}} \cosh(\tilde{\varphi}_j(x_1) - \pi\varepsilon_j) + \frac{Q_j(x_1)}{\sqrt{\Xi(x_1 - a)}} \sinh(\tilde{\varphi}_j(x_1) - \pi\varepsilon_j) \right), \quad x_1 \in L,$$

where

$$\tilde{\varphi}_j(x_1) = 2\varepsilon_j \arctan \sqrt{\frac{\Xi(b - c)\Xi(a - x_1)}{\Xi(a - c)\Xi(x_1 - b)}}, \quad x_1 \in L.$$

The intensity factors at the point b are defined by the following formulae:

$$K_{1b} = \lim_{x_1 \to b+0} \sqrt{2\pi \Xi(x_1 - b)}\sigma_{33}^{(1)}(x_1, 0), \quad K_{4b} = \lim_{x_1 \to b+0} \sqrt{2\pi \Xi(x_1 - b)}D_3^{(1)}(x_1, 0).$$
$$\tag{4.123}$$

Using expressions for the stresses and taking into account that $\lim_{x_1 \to b+0} \tilde{\varphi}_j(x_1) = \pi\varepsilon_j$, the following system of equations is obtained for the determination of the intensity factors:

$$\begin{cases} K_{1b} + m_{14}K_{4b} = \sqrt{2\pi}\dfrac{2e^{\pi\varepsilon_1}}{\sqrt{\Xi(b - c)}}P_1(b), \\ K_{1b} + m_{44}K_{4b} = \sqrt{2\pi}\dfrac{2}{\sqrt{\Xi(b - c)}}P_4(b), \end{cases}$$

solution of which is

$$K_{1b} = 2\sqrt{2\pi}\frac{m_{44}e^{\pi\varepsilon_1}P_1(b) - m_{14}P_4(b)}{(m_{44} - m_{14})\sqrt{\Xi(b - c)}}, \quad K_{4b} = 2\sqrt{2\pi}\frac{P_4(b) - e^{\pi\varepsilon_1}P_1(b)}{(m_{44} - m_{14})\sqrt{\Xi(b - c)}}.$$
$$\tag{4.124}$$

Here $P_1(b)$ and $P_4(b)$ have the following form:

$$P_1(b) = C_{11}\cos\frac{\pi(b - c)}{2h} + C_{21}\sin\frac{\pi(b - c)}{2h}, \quad P_4(b) = \sigma_4^*\sin\frac{\pi(b - c)}{2h}.$$

Using the condition $K_{1b} = 0$, the transcendental equation is obtained to determine the contact zone length

$$m_{44}e^{\pi\varepsilon_1}P_1(b) - m_{14}P_4(b) = 0. \tag{4.125}$$

Numerical results. The study of the periodic set of cracks is focused on the influence of the electric permittivity on the electric displacement D inside the cracks, the contact zone length and the intensity factors. The influence of the ratio of the crack length to the period, the ratio of the applied tensile and shear loading, the applied electric displacement on the electromechanical fields inside the composite is studied as well. The numerical results are presented for various combinations of piezoelectric (or piezoelectric and dielectric) materials.

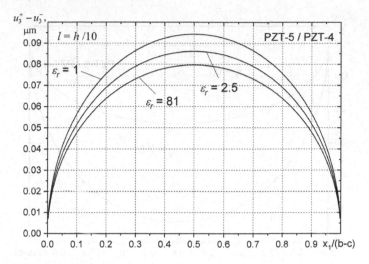

Fig. 4.25   Crack faces opening

Consider the influence of the external mechanical and electrical loading on the electric displacement inside the crack, the contact zone length and the SIF for some electric permeability values of the material inside the cracks such as air ($\varepsilon_r = 1$), silicon oil ($\varepsilon_r = 2.5$) and water ($\varepsilon_r = 81$). Values of ε_r equal to 10^{-6} and 4000 can be considered corresponding to the completely impermeable and permeable cracks. The period is taken π mm, the length of the crack is 0.1 of the period. The crack faces opening under the action of tensile mechanical loading $\sigma = 10$ MPa, shear mechanical loading $\tau = 0$ MPa and electric displacement $d = 0.001$ C/m^2 for different relative electric permeability coefficients is shown in Fig. 4.25. Piezoelectric bimaterial PZT-5/PZT-4 is chosen for the calculations.

The results for the electric potential jump across the interface are presented in Fig. 4.26 for the same geometrical parameters and loading as in Fig. 4.25.

Electric displacement distribution along the cracks according to Eq. (4.53) is obtained by the numerical integration of formulas (4.116) and (4.117). The results are illustrated in Fig. 4.27. It is important to note that the distribution of D_3 inside the cracks is close to the constant value defined by Eq. (4.96) with small deviation at the crack tips.

Application of the electric field regardless of its direction is confirmed to be restraining factor which leads to increase of the contact zone length and decrease of the SIF K_2. This conclusion agree with the study [13], in which the problem for a single crack of a finite electric permittivity in the homogeneous piezoelectric material was solved by the finite-element method, and with [11], in which the problem for a single crack in the piezoelectric bimaterial was solved analytically. When the coefficient of the electric permittivity ε_r is large, the results are similar to thoseobtained

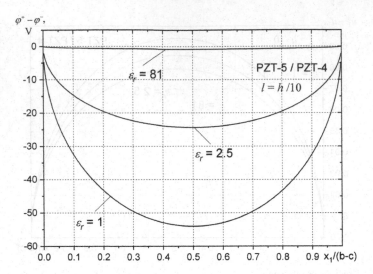

Fig. 4.26 Electric potential jump across the interface

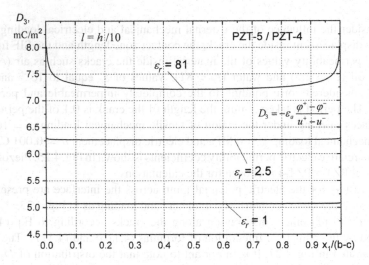

Fig. 4.27 Electric displacement field along the cracks

within the electrically permeable model. The value ε_r for air is equal to 1, the values 10^{-6} and 4000 can be considered corresponding to the electrically insulating and electrically permeable cracks, respectively.

Table 4.11 presents the results for the electric displacement D inside the cracks and the relative contact zone length λ.

The case of combined mechanical loading $\sigma = 10$ MPa, $\tau = -10$ MPa with different values of external electric displacement d and relative electric permeability of the cracks ε_r is investigated. The period is taken π, the length of the crack is taken

Table 4.11 Electric displacement D inside the cracks and relative contact zone length λ for the bimaterial PZT-PIC151/SiC

	$D, C/m^2$	$\lambda, 10^{-5}$
$d = 0.001\ C/m^2, \varepsilon_r = 1$	0.00097804	7.5966
$d = 0.01\ C/m^2, \varepsilon_r = 1$	0.00995519	7.4888
$d = 0.03\ C/m^2, \varepsilon_r = 1$	0.0299041	7.2510
$d = 0.001\ C/m^2, \varepsilon_r = 2.5$	0.00097953	7.6037
$d = 0.01\ C/m^2, \varepsilon_r = 2.5$	0.00997039	7.5604
$d = 0.03\ C/m^2, \varepsilon_r = 2.5$	0.029950	7.4645
$d = 0.001\ C/m^2, \varepsilon_r = 81$	0.00098049	7.6082
$d = 0.01\ C/m^2, \varepsilon_r = 81$	0.0099802	7.6069
$d = 0.03\ C/m^2, \varepsilon_r = 81$	0.0299796	7.6039

half of the period ($l = 0.5h$). Bimaterial consisting of the piezoceramics PZT-PIC151 and the semiconductor SiC is chosen for the numerical calculations.

Dependency of the relative contact zone length λ and normalized SIF K_2 on the applied electric displacement is presented in Fig. 4.28 for the constant mechanical loading $\sigma = 10\,MPa$, $\tau = -10\,MPa$. It can be easily seen that increase of the applied electric loading leads to increase of the relative contact zone length and decrease of the absolute value of the normalized SIF. This outcome underlines that application of the electrical loading enhance the crack growth resistance of the piezoelectric bimaterials.

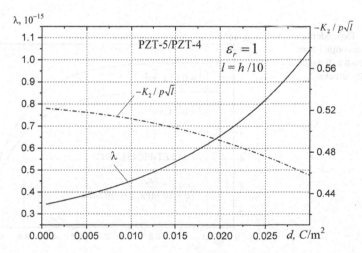

Fig. 4.28 Influence of the electric field on the relative contact zone length λ and normalized SIF K_2

It can be seen from the results of Table 4.11 and Fig. 4.28 that the influence of the external electrical loading is more significant with decreasing ε_r, because lesser electrical conductivity leads to greater electrical singularity in the vicinity of the crack tips. For the small values of the electric displacement, the obtained IFs for any electric permittivity values are in a good agreement with the corresponding results for the completely permeable crack.

In Fig. 4.29, the dependency of the relative contact zone length on the ratio of the crack length to the period is illustrated for the bimaterial PZT-PIC151/SiC. Constant electrical loading and two combinations of the tensile and shear mechanical loading are considered. Electric permittivity corresponds to the cracks filled with air. It can be easily seen that for a periodic case the crack interaction becomes essential when the crack length exceeds half of the period, i.e. when the crack length is larger than the distance between the cracks. As expected, SIF K_2 increases with increasing the ratio of the crack length to the period.

Conclusions. A problem for the periodic set of cracks is solved within the contact model taking into account their electric permittivity. The expressions for the mechanical stresses, elastic displacements, electric potential and electric displacements are obtained at the interface of the materials. The equation for the determination of the electric displacement inside the cracks is derived. The transcendental equation for the contact zone length is derived as well. The expressions for the mechanical and electrical intensity factors are obtained. It is found that the contact zone length and the SIF depend both on the electrical permeability of the crack medium and on the intensity of the applied electric displacement. It follows from the numerical analysis of the solution that application of the electric field restrains the development of the cracks. This can be seen from the increase of the contact zone and decrease of the IF.

Fig. 4.29 Dependence of the relative contact zone length λ on the distance between the cracks

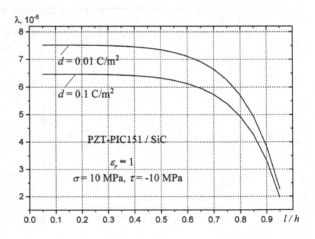

4.5 Conclusions

In this section a solution for the periodic set of cracks located at the interface of two dissimilar piezoelectric materials is obtained for the model of electrically permeable crack as well as for the more realistic model of finite electric permittivity of the crack filler. Both models are considered within the classical and contact approaches. Detailed analysis of the results and conclusions are presented at the end of each subsection.

References

1. Suo Z, Kuo CM, Barnett DM, Willis JR (1992) Fracture mechanics for piezoelectric ceramics. J Mech Phys Solids 40, 739–765 (1992)
2. Gakhov F (1966) Boundary value problems. Pergamon Press, Oxford
3. Parton V, Kudryavtsev B (1988) Electromagnetoelasticity. Gordon and Breach Science Publishers, New York
4. Kozinov S, Loboda V (2010) A periodic system of electrically permeable cracks at the interface between two piezoelectric materials. J Math Sci 170(5):659–673
5. Kozinov S, Loboda V, Lapusta Y (2013) Periodic set of limited electrically permeable interface cracks with contact zones. Mech Res Commun 48:32–41
6. Herrmann K, Loboda V (2000) Fracture mechanical assessment of electrically permeable interface cracks in piezoelectric bimaterials by consideration of various contact zone models. Arch Appl Mech 70:127–143
7. Kozinov S, Sheveleva A, Loboda V (2019) Fracture behavior of periodically bonded interface of piezoelectric bi-material under compressive-shear loading. Math Mech Solids 24(10):3216–3230
8. Sheveleva A, Lapusta Y, Loboda V (2015) Opening and contact zones of an interface crack in a piezoelectric bimaterial under combined compressive-shear loading. Mech Res Commun 63:6–12
9. Hao T, Shen Z (1994) A new electric boundary condition of electric fracture mechanics and its application. Eng Fract Mech 47:793–802
10. Loboda V, Kozinov S (2011) Periodic set of the interface cracks with limited electric permeability. In: Book of proceedings of the international IUTAM symposium 'Multiscale modelling of fatigue, damage and fracture in smart materials systems', Freiberg, Germany, vol 24 (2011)
11. Govorukha V, Loboda V, Kamlah M (2006) On the influence of the electric permeability on an interface crack in a piezoelectric bimaterial compound. Int J Solid Struct 43:1979–1990
12. Li Q, Chen Y (2008) Solution for a semi-permeable interface crack in elastic dielectric/piezoelectric bimaterials. ASME J Appl Mech 75:1–13
13. Gruebner O, Kamlah M, Munz D (2003) Finite element analysis of cracks in piezoelectric materials taking into account the permittivity of the crack medium. Eng Fract Mech 70:1399–1413

Conclusions

In the presented monograph the analytical methods are developed for the plane problems of electroelasticity for composites, consisting of two isotropic, anisotropic or piezoelectric materials with the periodic or arbitrary set of cracks at the interface of the materials.

As a result of the research the following main results are derived:

- formulation of the linear conjunction problem and obtaining of the closed-form solution for the periodic set of cracks located at the interface of dissimilar isotropic or anisotropic materials with smooth contact zones of their faces subjected to the mechanical loading. This allowed to establish the dependence of the contact zone length and the SIFs on the distance between the cracks, mechanical parameters of the bimaterial and the external loading. On the basis of the obtained results the following conclusions are drawn: the crack growth resistance of a bimaterial decreases with approaching cracks, with increasing the differences of elastic properties of the components, with decreasing the anisotropy value and almost does not depend on the inclination angle of the external loading.
- a new closed-form solution for an arbitrary set of the interface cracks with contact zones of their faces in isotropic and anisotropic bimaterials at the remote tensile-shear loading is obtained. This solution allows to investigate the dependence of contact zones and the SIFs on the relative position of the cracks and their relative lengths. In case of two cracks it is shown that the crack growth resistance of a bimaterial decreases with approaching cracks and with increasing difference in their lengths. The cracks will propagate towards each other.
- a new solution to the elastic plane problem for a periodic set of the fully electrically permeable cracks, located along the interface of two electrically active materials is obtained. This solution illustrates the variation of the contact zones and IFs depending on the distance between cracks and the inclination angle of an external loading. As a limiting case of the solution for the contact model, a periodic set

S. Kozinov and V. Loboda, *Fracture Mechanics of Electrically Passive and Active Composites with Periodic Cracking along the Interface*, Springer Tracts in Mechanical Engineering, https://doi.org/10.1007/978-3-030-43138-9

of the electrically permeable cracks is studied within the classical "oscillating" model. The stress fields obtained within the oscillating and contact models are compared along the interface.

– a solution for the periodic set of internal cracks in a homogeneous piezoelectric space is found taking into account a finite electric permittivity of the crack filler. Explicit formulas for the stresses and electric displacement, mechanical and electrical IFs and the energy release rate are found. It is established that for the cracks with a finite electric permittivity the ERR varies disproportionately under the linear change of the external electromechanical loading. The results under low electrical loading are similar to those obtained for the electrically permeable cracks.

– a periodic set of cracks located along the interface of two piezoelectric materials is investigated taking into account a finite electric permittivity of the crack filler. Both the oscillating and contact models are considered. New analytical formulae for the stresses and electric displacement, mechanical and electrical IFs in the composite are established. The magnitude of the electric displacement inside the cracks is found. The influence of the electric permittivity of cracks and the ratio of their lengths to the period on the main fracture parameters is shown. It is established that, regardless of its direction, the applied electrical loading retards the development of the cracks.

Printed in the United States
by Baker & Taylor Publisher Services

Printed in the United States
by Baker & Taylor Publisher Services